IMAGES
of America

LAKE ERIE'S SHORES AND ISLANDS

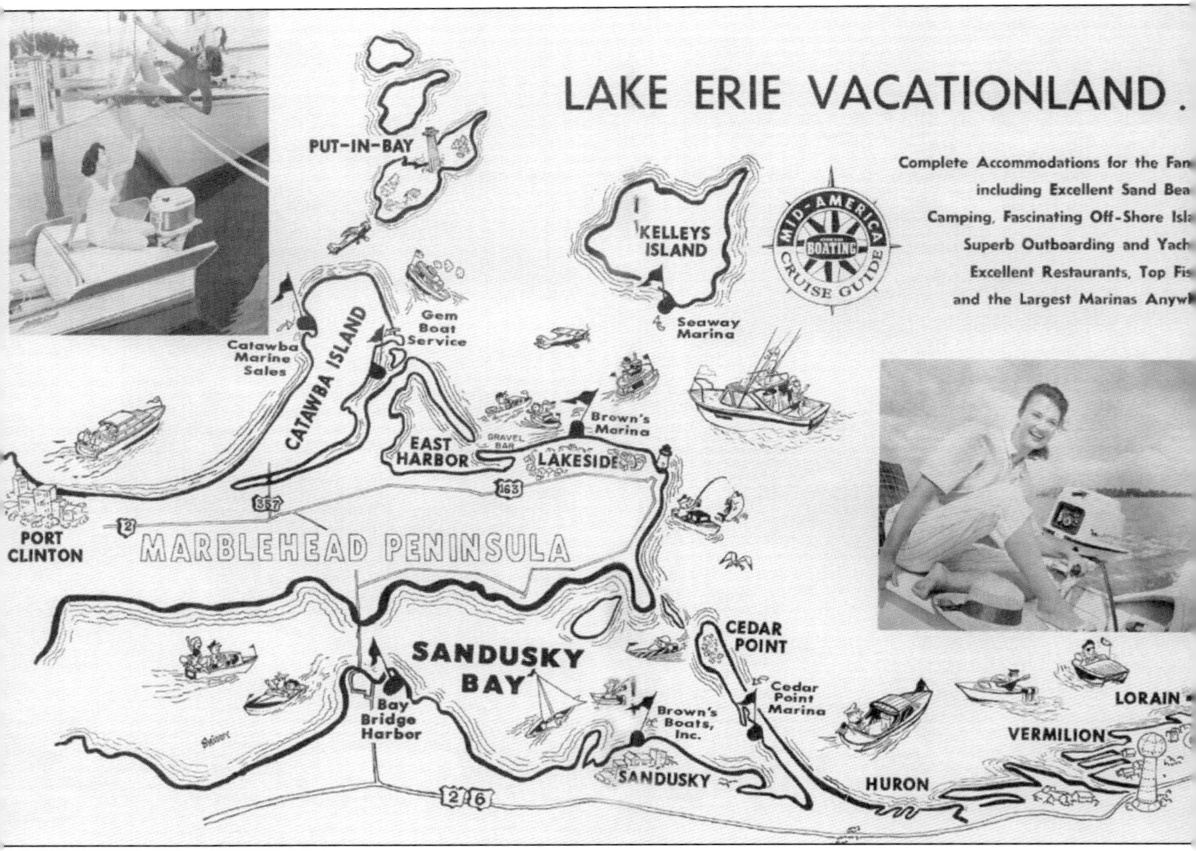

Over the decades, a number of colorful and whimsical souvenir maps have been produced for the shores and islands region of Ohio. This map appeared in a special advertising supplement in the June 1959 issue of *Ohio Boating*. For reference, the distance from Vermilion to Port Clinton via US Route 6 is 38 miles. (Courtesy of Frank Landl.)

On the Cover: Bathers are seen on the Cedar Point Beach around 1910. The new bathhouse and the Coliseum are in the background. (Courtesy of the Hayes Presidential Center, Fremont, Ohio.)

IMAGES
of America

LAKE ERIE'S SHORES AND ISLANDS

H. John and Marie Hildebrandt

ARCADIA
PUBLISHING

Copyright © 2015 by H. John and Marie Hildebrandt
ISBN 978-1-4671-1372-4

Published by Arcadia Publishing
Charleston, South Carolina

Printed in the United States of America

Library of Congress Control Number: 2014950914

For all general information, please contact Arcadia Publishing:
Telephone 843-853-2070
Fax 843-853-0044
E-mail sales@arcadiapublishing.com
For customer service and orders:
Toll-Free 1-888-313-2665

Visit us on the Internet at www.arcadiapublishing.com

This book is dedicated to Joan Van Offeren, longtime friend, colleague, and true daughter of Lake Erie's shores and islands.

Contents

ACKNOWLEDGMENTS

We are grateful to a number of organizations and individuals who provided assistance in creating *Lake Erie's Shores and Islands*. Nan Card at the Hayes Presidential Center helped us review the Charles Frohman, Ernst Niebergall, and Reverend Henry Cooke collections and also provided counsel throughout the project. Julie Mayle at Hayes was also very helpful. Ron Davidson and Maggie Marconi of the Sandusky Library Archives Research Center and the Follett House Museum were of great assistance. Brenna Walker at the Sandusky Area Maritime Association helped us locate images related to the region's maritime history.

Joan Van Offeren and Dawn Weinhardt of the Lake Erie Shores and Islands Visitors Bureau were helpful in countless ways.

Ann Marie Muehlhauser, Barb Wozniak, John Taylor, and Lee Alexakos at Cedar Point helped us review the park's collection of archival images. Jillian Carney and Lily Birkimer at the Ohio History Connection were of great assistance in working with the OHC archives. Ray Parker at Lyme Village, Kurri Lewis at the Merry-Go-Round Museum, Gail Alt at Sorrowful Mother Shrine, Lance Woodworth at the Jet Express, Tim Tester at Mad River Railroad, and Jack Coffman and Paul Schoenegge in Castalia were also helpful.

A special thank you to Zack Dolyk, who unlocked the key to Vermilion, and to the Vermilion Archival Society, including Margaret Wakefield Worcester and Bob Reese. Leslie Korenko was helpful in obtaining images of Kelleys Island. Mark Ackerman of Hart Advertising located vintage billboard art of the area.

A number of individuals shared their family photographs and postcard collections with us: Diane Francis, Ben and Melissa Martilotta, Marlene and Ed Boas, Sheila Ehrhardt, Kula Hoty Lynch, Barb Falfas, Frank Landl, James Hildebrandt, Denise Bell, Kitty Smith, the late Marje Rody, and Jim Lang.

The authors would also like to thank Lisa Crescimano, Eric Pasley, and Shelley Bryant, who shepherded us through several technical twists in the project.

The following abbreviations are used in photograph credits: Cedar Point Archives (CPA), Hayes Presidential Center (HPC), Lake Erie Shores and Islands Visitors Bureau (LESI), Sandusky Area Maritime Association (SAMA), and Sandusky Library Archives Research Center (SLARC).

INTRODUCTION

From the observation deck on Perry's Monument, 317 feet above Put-in-Bay Harbor, the view to the east reveals Kelleys Island, then nothing but a blue horizon leading off to Buffalo, approximately 210 lake miles away. To the south is the soft green line of the Marblehead Peninsula. To the west is a necklace of small islands: Rattlesnake, Green, Sugar, and East Sister. If it is clear, Monroe, Michigan, and the mouth of the Detroit River can be seen. To the north are Middle and North Bass Islands, and just seven miles away is the southernmost point of the second largest country in the world: Middle Island, Canada. Below the observation deck is the village of Put-in-Bay, its harbor filled with boats, and Gibraltar Island, the turret of Cooke's Castle sticking up through the trees. The vista is dazzling, the equal of any on the Great Lakes.

A short 14 miles southeast, from the summit of one of Cedar Point's majestic roller coasters, the view is equally impressive: blue water, sky, green-robed islands, specks of boats and their trailing wakes, and the invisible line that separates Canada from the United States.

For 150 years, this region of Ohio has had a special allure for visitors. The area from Vermilion on the east to Port Clinton on the west, south to Milan, Bellevue, and Fremont, north to include Sandusky, Marblehead, Catawba, and the Lake Erie islands has been labeled "Ohio's Vacationland," and "Lake Erie Land," and now "Lake Erie Shores and Islands."

For Myrtle Beach, the attraction is the beach. For the Poconos, it is the mountains. Here, it is Lake Erie, the natural resource that gives the area its character and charm. The shores and islands area is approximately 1,250 square miles, 60 percent of which is water.

Beginning in the 1860s, visitors came to the islands of South Bass and Middle Bass by steamer from Detroit and Toledo to sample island wines and enjoy the natural beauty of the islands. The climate favored grape growing (and still does), and a large wine industry grew up on the islands and on the mainland.

Jay Cooke, Civil War financier and Sandusky native, was one of the first to recognize the appeal of the Lake Erie islands. In 1864, he bought Gibraltar Island, which forms the northern boundary of Put-in-Bay Harbor, and built a magnificent summer home called Cooke's Castle. For the next 50 years, his family spent summers on the island. It is now owned by The Ohio State University, where it operates Stone Laboratory for Great Lakes Research.

Cedar Point traces its first season as a resort to 1870, when the steamer *Young Reindeer* offered trips to residents to swim and picnic on the beach. Visitors loved the natural beauty of the Cedar Point Peninsula; within a few years, the owners of the property began adding amusements: band concerts, dancing, boat rentals, and food concessions. In 1888, they opened the Grand Pavilion on the Cedar Point beach, which housed a cavernous main hall, bowling alleys, a photography studio, a large barroom, a fully equipped kitchen, and even three gas-operated chandeliers. Although much modified, the building is the oldest existing structure at Cedar Point and has operated for parts of three centuries.

Who were the visitors to Put-In-Bay and Cedar Point in these early years? Geographically, they were mostly residents of Ohio, Michigan, and surrounding states and the fast-growing cities of Cleveland, Detroit, Toledo, Akron, Pittsburgh, Youngstown, and Columbus. Demographically, they represented the growing American middle class, who in the years after the Civil War enjoyed more time for leisure activities.

In the years leading up to the turn of the century, South Bass and Middle Bass Islands enjoyed a golden age of tourism development. Many of the hotels and resorts on the islands catered to the upper class. In 1874, a group of businessmen from the Toledo area built a "clubhouse" on Middle Bass to create a summer retreat. Several families built large summer homes on the island. The "Middle Bass Club" offered bowling, tennis, boating, and weekend bands for dancing.

The Hotel Victory, built in 1892 on a bluff on the west side of South Bass Island, staked a claim as the largest summer hotel in the United States. It had 625 rooms (including 85 with private baths). The main dining room was 155 feet by 85 feet, and the hotel could feed up to 1,250 guests in one sitting. Located a mile and a half from the steamboat docks at Put-In-Bay, it was served by an electric streetcar line. Never a financial success, the Hotel Victory burned to the ground on the night of August 14, 1919.

Transportation played a key role in tourism development in the region. Steamships carried passengers and cargo to and from the islands and the ports along the Lake Erie shores for nearly a century, starting in the 1850s and 1860s. The last steamships were scrapped after World War II. In the summer, the harbors of Put-in-Bay and Sandusky were filled with Lake Erie steamers, connecting Sandusky, Cedar Point, Lakeside, the islands, Toledo, Detroit, and Cleveland. During this period, photographs taken of the piers in Sandusky or Put-in-Bay often show three or more steamers in the process of boarding passengers. As mud-filled roads were upgraded and new infrastructure added, including the Sandusky Bay Bridge in 1929, the shores and islands region also became more accessible via automobile.

Everything changed for Cedar Point when it was purchased by Indiana entrepreneur George A. Boeckling in 1899. The years leading up to the turn of the century had been lean ones for Cedar Point, but Boeckling was convinced that the resort had unlimited potential. In less than 10 years, he turned Cedar Point into a thriving amusement park as well as a resort, adding a midway with rides and building the 600-room Hotel Breakers in 1905 and the Coliseum in 1906. Boeckling was a very effective promoter. He brought in vaudeville acts and brass bands, staged championship boxing matches, and even sponsored the Glenn Curtiss flight from Euclid Beach to Cedar Point—at 65 miles, the first long-distance flight over water. Cedar Point thrived under Boeckling's energetic leadership. The resort continued to grow and prosper through the 1920s.

Cedar Point was not the only amusement park in the region. Crystal Beach Park, on the east side of Vermilion, opened in 1907 and operated until 1965. Named for the crystal-like quality of its beach sand, it was known as the park with a thousand trees.

The shores and islands area also became home to a number of summer communities with a religious orientation. The largest and best known is Lakeside, founded in 1873 by the local Methodist Church, which had a dream for a camp meeting location and a summer resort for Christian families. The Hotel Lakeside opened in 1875 and is still in service. In its early years, Pres. Rutherford B. Hayes was a frequent visitor. Lakeside has grown into the "Chautauqua on Lake Erie." Other religious communities staked out a piece of the Lake Erie shore to create summer retreats, including Linwood Park, Beulah Beach, and Heidelberg Beach in Vermilion.

The Depression years were hard on the tourism industry, reflecting economic conditions throughout the United States. Development at Cedar Point, as well as on the islands, came to a halt. Maintaining, even surviving, was the goal of most businesses. With Boeckling's death in 1931, Cedar Point slipped into the doldrums for 25 years. Most wineries, shuttered through Prohibition, reopened, but to a smaller market.

The postwar economic recovery and the growing number of baby boom families were a boost to the region. Car travel was easier, and families had more leisure time and money. Cedar Point got a new lease on life when it was acquired in 1956 by a group headed by George Roose of Toledo, a bond dealer, and Emile Legros, a Cleveland banker. Although they purchased the property with the intention of razing it to build a housing development, Roose and Legros were dissuaded by citizen groups—and the State of Ohio—and decided to give it a try as an amusement park.

In 1960, armed with a study that indicated the marketplace would respond to the right investment, the two men borrowed $16 million and began to revitalize Cedar Point as an amusement park and resort, adding a marina, a new midway, and new rides and attractions. Perhaps as important, the men committed to a guest-centered operating philosophy, using Disneyland as a model. Attendance increased dramatically, reaching two million in 1965, and Cedar Point never looked back, adding "something new" every season. In 1970, when it celebrated its centennial, Cedar Point had been transformed and was poised for a new golden age.

During the same period, industrious men and women saw the potential for small business development in the wake of Cedar Point's success and began opening motels, restaurants, stores, and attractions to accommodate the new visitors, including a growing number of overnight visitors.

In the 1970s and 1980s, Lake Erie experienced a renaissance. The pollution and algae problems of the 1960s—which had reached a climax with the infamous fire in 1969 on the Cuyahoga River in Cleveland—were abated thanks to reduced phosphorus flow into the lake and improvement in waste treatment facilities. By the late 1970s, the lake was dramatically improving; with improved lake conditions came the return of the walleye and yellow perch. One statistic tells the story: in 1975, there were 34 charter boat captains working in western Lake Erie. In the mid-1980s, there were more than 1,200. The return of the walleye fishery spurred economic activity, including bait shops, boat sales, retail sales, cottage rentals, hotel occupancy, restaurant meals, and fuel sales.

Along with better fishing came an overall change in perception regarding Lake Erie. It was cool again. At first, the good word was confined to the immediate area, but soon it spread throughout Ohio and neighboring states. One benefit of the change in perception was the growth of condominium and second-home communities, especially those close to the lake in Sandusky, Marblehead, and Catawba. These developments were a far cry from the small cottages and trailers that served as second homes for generations. Many developments targeted an upscale consumer. Le Marin on Catawba Island and The Harbour on Sandusky Bay both opened in the mid-1980s and compared favorably to similar developments in Florida and the Carolinas.

Cedar Point continued to grow through the 1970s and 1980s. Attendance topped three million for the first time with the introduction of the revolutionary Corkscrew (the first triple-looping roller coaster) in 1976. The park followed up two years later with Gemini, the highest, steepest, and fastest roller coaster in the world when it debuted on June 17, 1978. Cedar Point's success, as well as the popularity of sportfishing and boating, led to the development of many new hotels in the area, as well as new marinas, restaurants, and vacation home developments.

The past 25 years have seen continued tourism growth in the shores and islands region, driven by three main factors: the addition of world-class roller coasters and new resort development at Cedar Point; the introduction of a new type of attraction, the indoor water park; and the development of a new transportation system, the Jet Express, to take visitors to Put-in-Bay and the other islands.

Under the leadership of CEO Dick Kinzel, Cedar Point introduced a succession of world-class coasters during this period, starting with Magnum in 1989 and followed by Mean Streak (1991), Raptor (1994), Mantis (1996), Millennium Force (2000), Wicked Twister (2002), Top Thrill Dragster (2003), and Maverick (2007). Following Kinzel's retirement in 2011, new CEO Matt Ouimet continued the coaster tradition with GateKeeper in 2013, creating a coaster lineup unmatched by any park in the world. At the same time, Cedar Point went back to its roots as a resort by adding new hotel properties and refurbishing the venerable Hotel Breakers. Cedar Point has successfully positioned itself as a vacation destination as well as an amusement park.

Based on a concept developed in the Wisconsin Dells, a new attraction made its debut in 2001 in Sandusky. The 271-room Great Wolf Lodge is an indoor water park resort, combining a water park with a highly themed, family friendly hotel. It was an instant success and a year-round tourism draw. Until Great Wolf, winter tourism had been limited to ice boating and ice fishing (always problematic and with specialized appeal). Cedar Point's Castaway Bay followed in 2004. In 2005, Kalahari opened. After several additions, Kalahari now totals 173,000 square feet of water park and has 800 hotel rooms, plus restaurants, a spa, and more than 200,000 square feet of meeting and exhibit space. There are now five indoor water parks in the area, including Rain and Maui Sands.

The Jet Express made its maiden voyage from Port Clinton to Put-in-Bay in June 1989, helping to create a second golden age for Put-in-Bay. The Jet delivered guests from downtown Port Clinton to Put-in-Bay Harbor in just 20 minutes versus the traditional 90 minutes. Island visitors could travel faster—and stay later. This led to an increase in visitors, which in turn led to a boom in the development of restaurants, gift shops, hotels, bars, and other attractions.

Ecotourism has also developed in the shores and islands area during the past 25 years. Bird-watching has attracted many visitors to Kelleys Island, Sheldon's Marsh, Old Woman's Creek, East Harbor State Park, and Magee Marsh and the Ottawa National Wildlife Refuge. There are 24 public nature preserves, state parks, and wildlife refuges in Erie and Ottawa Counties.

Sandusky is now home to several quality museums: the Merry-Go-Round Museum, the Sandusky Area Maritime Museum, and the Follett House Museum. The area's historical attractions, including the Thomas Edison home in Milan, the Confederate Cemetery on Johnson's Island, the Marblehead Lighthouse (the oldest on the Great Lakes), the Perry's Victory and International Peace Memorial on South Bass Island, and the new Liberty Aviation Museum, attract many visitors. Sandusky is also a growing heritage tourism site for the Underground Railroad. Annual visitation to the shores and islands area now exceeds seven million people.

For 150 years, this area has attracted visitors in search of fun and relaxation. Lake Erie brings them here, and Lake Erie brings them back. Enjoy the story.

One

The Gilded Age Discovers Lake Erie 1860–1890

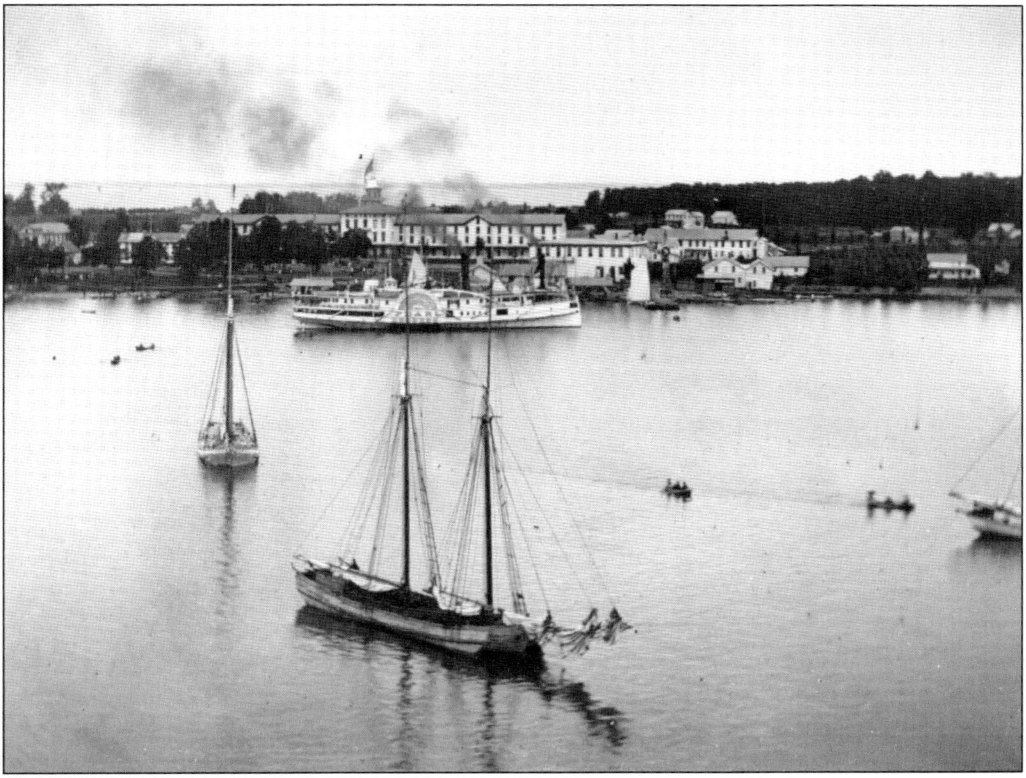

Western Lake Erie and the surrounding shores and islands first began attracting leisure visitors in the 1860s. The lure of the lake, natural harbors, and wineries contributed to the popularity of the area. Put-in-Bay became the focal point for excursions to the islands. This photograph of Put-in-Bay Harbor, taken from Gibraltar Island around 1880, shows the steamer *Pearl* docked near the Put-in-Bay House, a popular hotel. The land on the horizon is the Marblehead Peninsula. (Courtesy of HPC.)

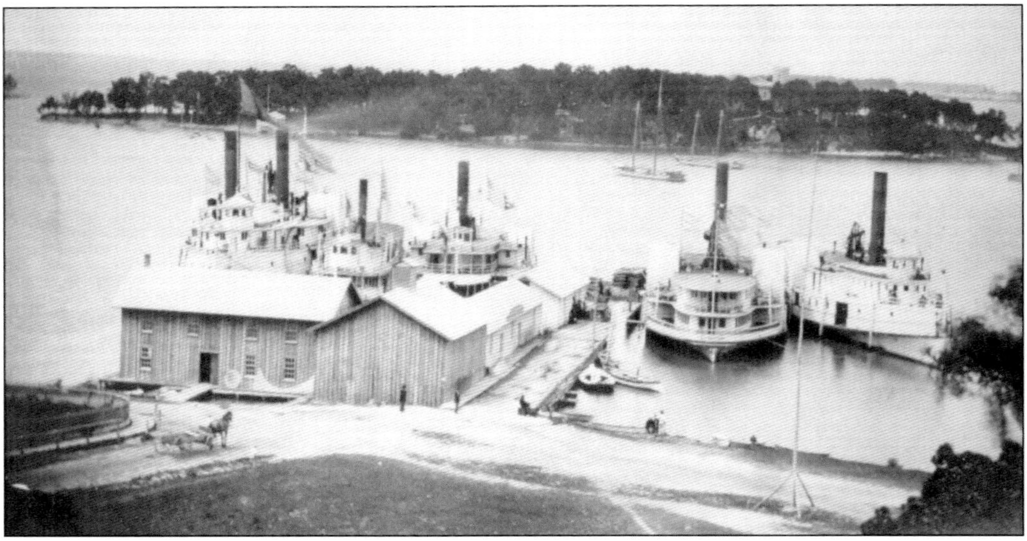

This c. 1880 photograph shows Put-in-Bay Harbor with Gibraltar Island in the background. The image was made from a series of stereographic cards created by A.C. Platt. By the early 1880s, the islands had become a very popular destination for daily excursions from cities like Detroit, Cleveland, and Toledo. (Courtesy of SLARC.)

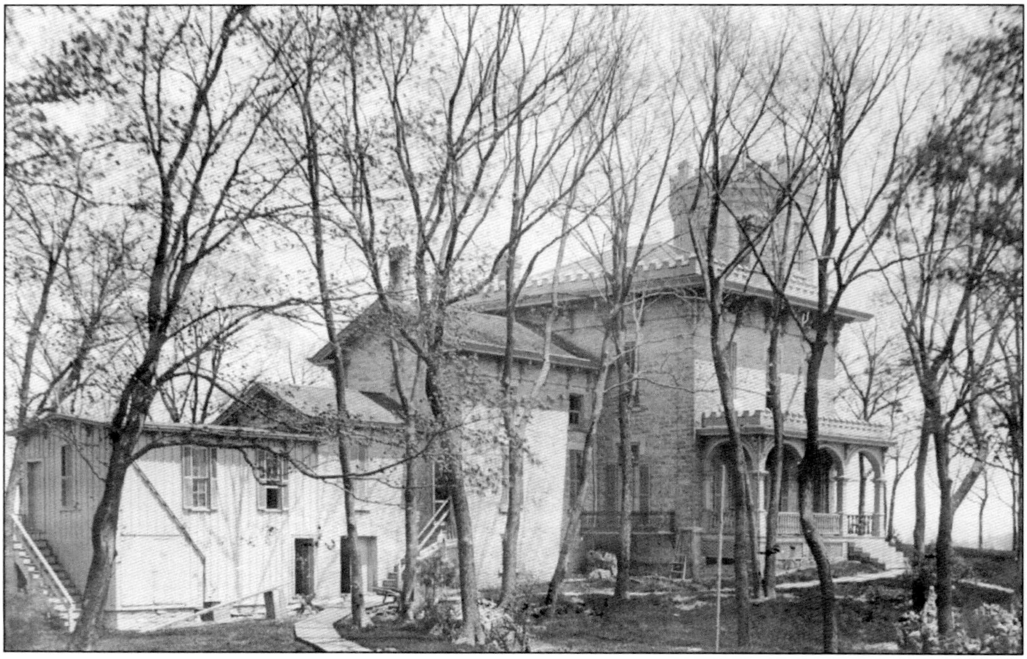

Jay Cooke, a native of nearby Sandusky, purchased Gibraltar Island from local landowner Joseph de Rivera St. Jurgo in 1864 for $3,000. He immediately began building a summer retreat, a beautiful 15-room limestone residence popularly known as "Cooke's Castle." Cooke was one of the richest men in America at the time, a financier who had successfully sold bonds to finance the Union effort during the Civil War. For more than 50 years, Cooke and his extended family journeyed from their main residences in Philadelphia to Gibraltar to enjoy Lake Erie summers. Cooke Castle was named a National Historic Landmark in 1966. The island has been owned by The Ohio State University since 1925. (Courtesy of HPC.)

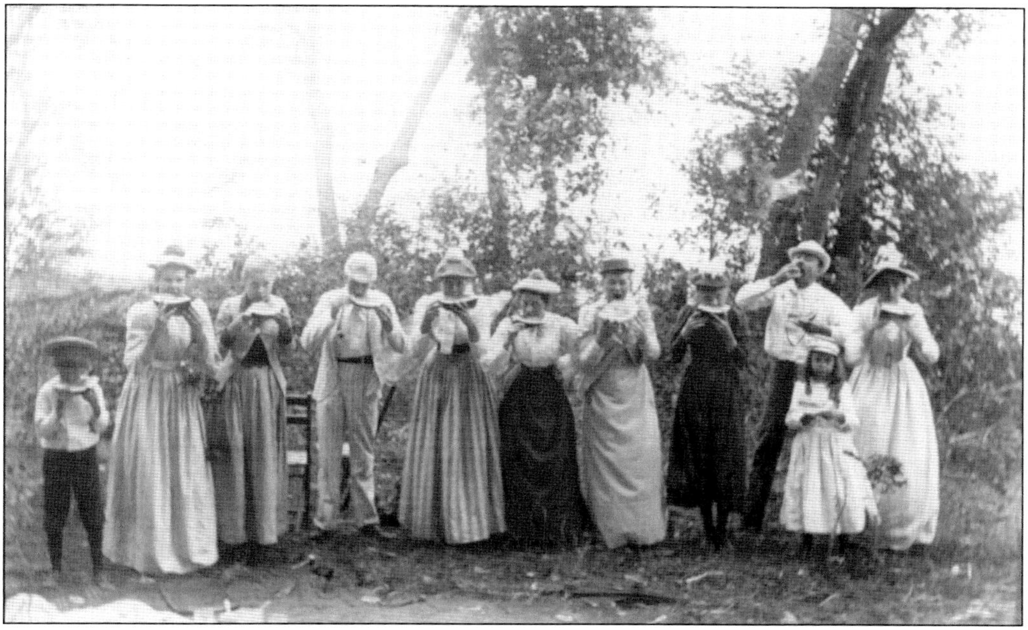

Jay Cooke's extended family enjoys a picnic on nearby Green Island around 1890. Watermelon is one of the menu items. Picnicking was a popular summertime activity in the 19th century, more so than today. Picnics often featured elaborate meals and were enjoyed at a pace that would take up most of the day. (Courtesy of HPC.)

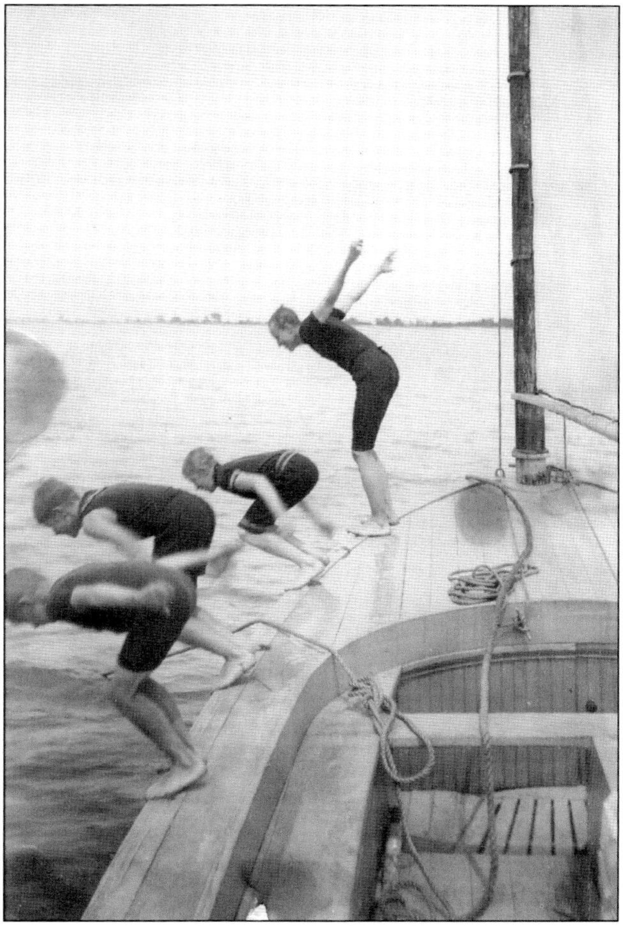

Cousins Jay Cooke III and Allen, John, and Henry Butler prepare to dive into the cool waters of Lake Erie on a warm day in the summer of 1889. The photograph was taken by Jay Cooke's son, the Reverend Henry Cooke, whose hobby was photography. For four decades, he recorded family activities on Gibraltar and the other islands in a remarkable collection of more than 2,000 images. (Courtesy of HPC.)

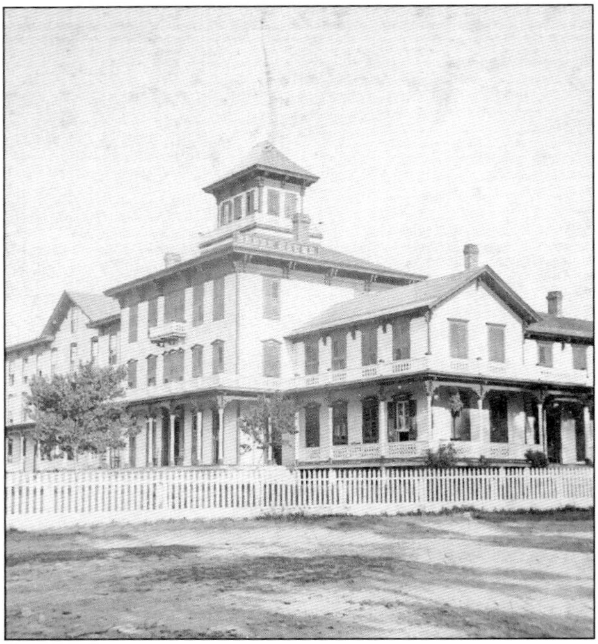

One of the first hotels on South Bass Island was the Beebe House, built in the early 1860s and expanded in 1869 by Henry Beebe. Located on Main Street in Put-in-Bay, the hotel, seen here around 1880, could accommodate 400 guests and offered a billiard room, bowling alley, and dancing hall. Rates were $14 per week, $2.50 per day, and half price for children and servants. (Courtesy of HPC.)

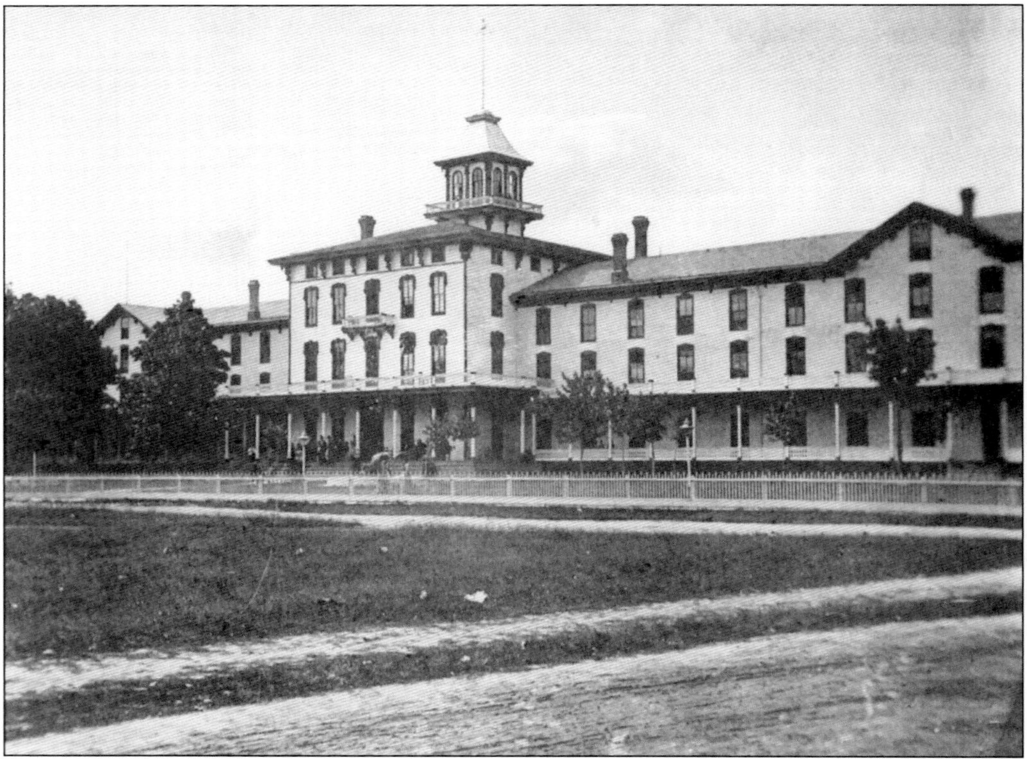

The Put-in-Bay House, shown here about 1875, was built in 1836 and opened as a rooming house in 1861. The hotel featured a 500-foot-long veranda and included a gymnasium, dance hall, and barbershop in adjacent buildings. The hotel burned to the ground in 1878, a common fate of wood-frame hotels of the period. A second Put-in-Bay House was constructed in 1890 and operated until 1907, when it suffered the same fiery fate as its predecessor. (Courtesy of HPC.)

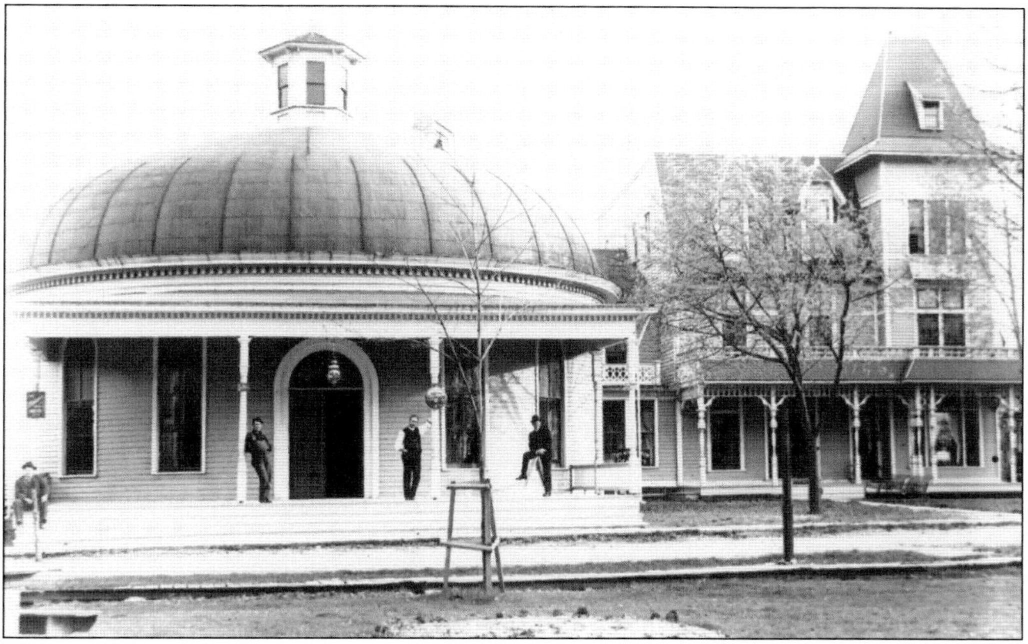

The Round House has been a prominent part of the Put-in-Bay landscape for more than a century. Originally built as a residence in Toledo, the building was dismantled and rebuilt at the corner of Delaware and Lorain Avenues, where it opened as the Columbia Restaurant in 1873. Today, it is one of the most popular bars on the island. The Park Hotel is at right in the c. 1890 photograph. (Courtesy of HPC.)

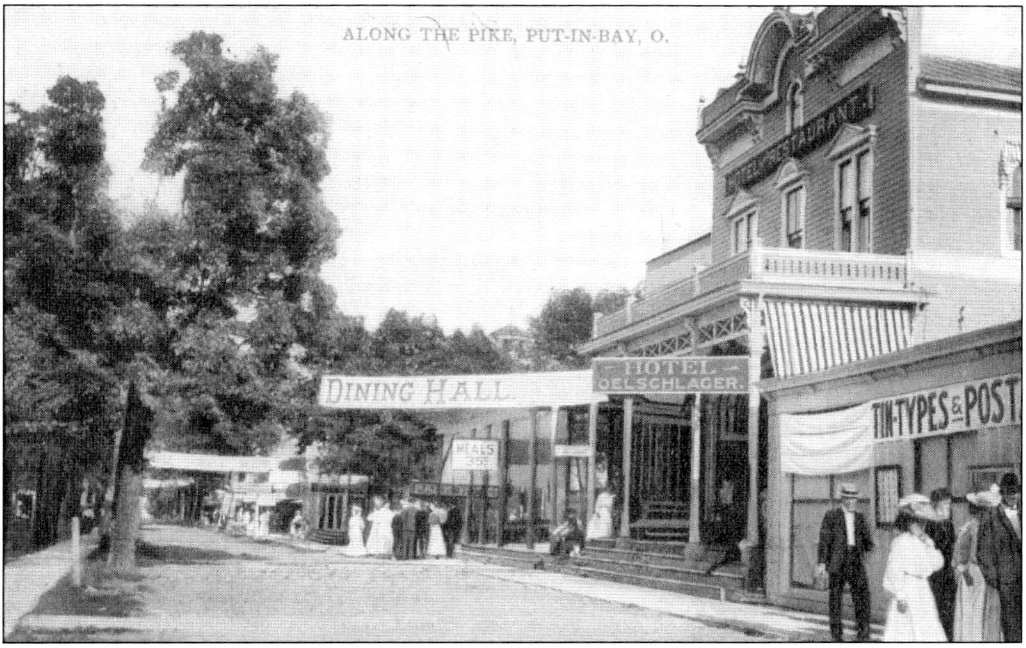

Delaware Avenue was the main commercial street in Put-in-Bay and a gathering place for visitors. The Hunker House, built in 1871, was known by several names during its long history. It was the Crescent Hotel when it closed in 1971. Here, guests gather in front of the hotel around 1890. (Courtesy of HPC.)

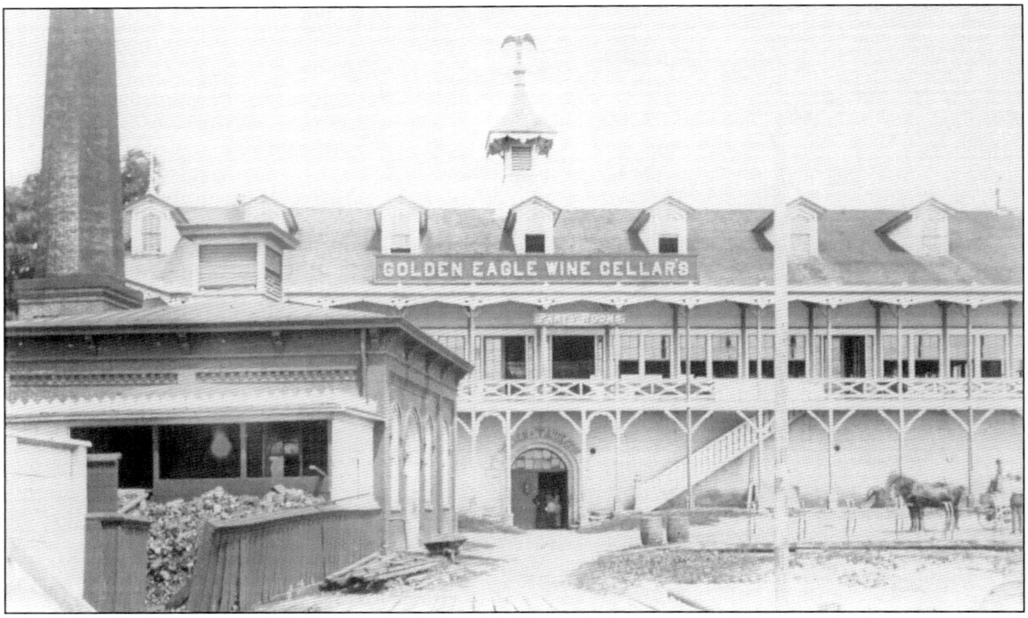

Grape cultivation was a mainstay on many of the Lake Erie islands, starting in the 1850s. German immigrants, mostly from the Rhineland area, tended the vineyards on Middle Bass and South Bass Islands. In the 1860s, a German settler named Andrew Wehrle built a winery at the southern tip of Middle Bass. A stonecutter by trade, he dug a 14-foot-deep wine cellar out of native limestone. The winery prospered, and by 1875 Golden Eagle Winery was reputed to be the largest winery in the United States. It is seen here around 1885. (Courtesy of SAMA.)

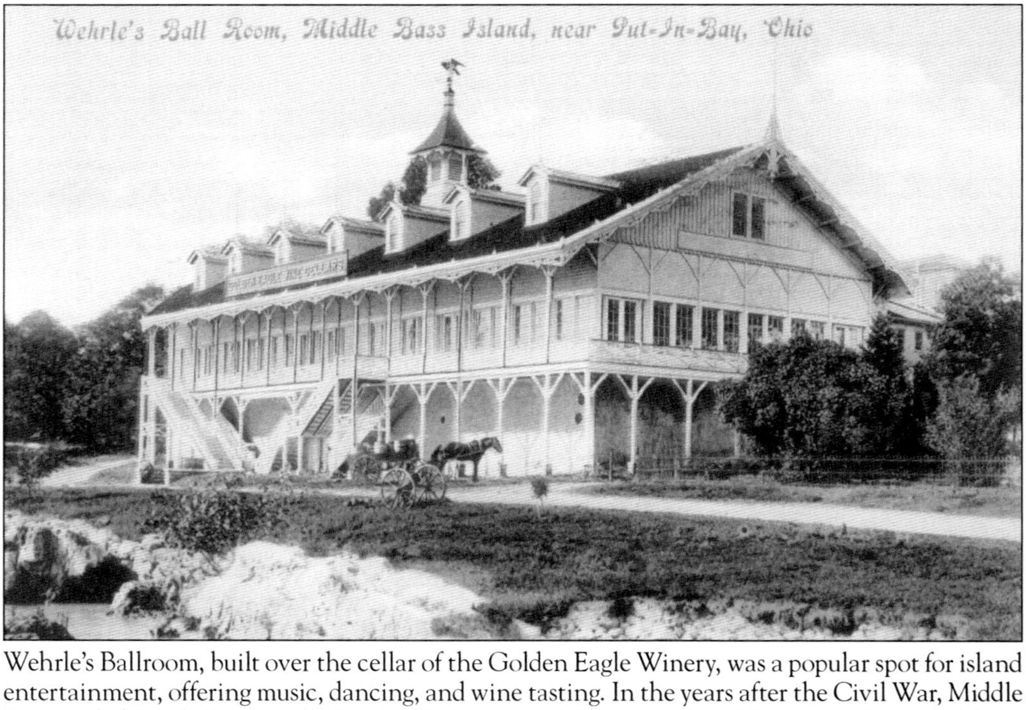

Wehrle's Ballroom, built over the cellar of the Golden Eagle Winery, was a popular spot for island entertainment, offering music, dancing, and wine tasting. In the years after the Civil War, Middle Bass rivaled South Bass as a destination for leisure activity, attracting visitors from Detroit, Toledo, Lorain, and Cleveland. This photograph was taken around 1888. (Courtesy of SLARC.)

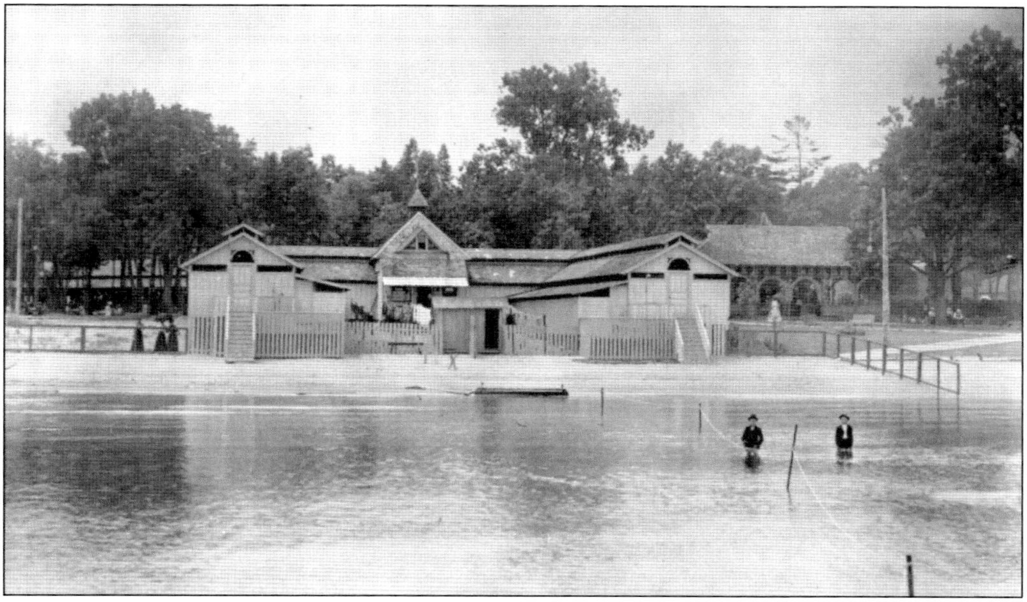

While the Lake Erie islands enjoyed an earlier popularity and drew visitors from longer distances, Cedar Point was also recognized as a popular and appealing destination to escape summer heat and humidity, especially among residents. Steamship service started in the summer of 1870, transporting visitors to the Cedar Point beach to swim and picnic. Cedar Point's first bathhouse is seen here in 1888. (Courtesy of CPA.)

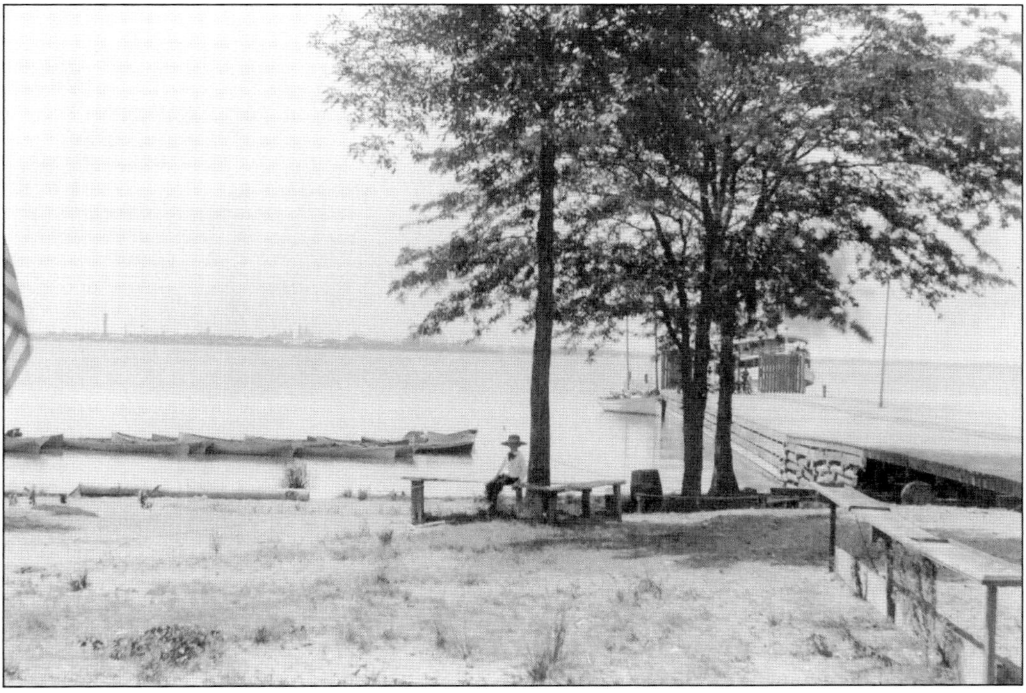

A young boy sits under a tree on the bay side of Cedar Point around 1885. He is likely responsible for managing the rental of the rowboats lined up at the water's edge. The city of Sandusky is a mile across the bay behind him. The steamer at the dock is unknown. Note the American flag in front of the boats. (Courtesy of HPC.)

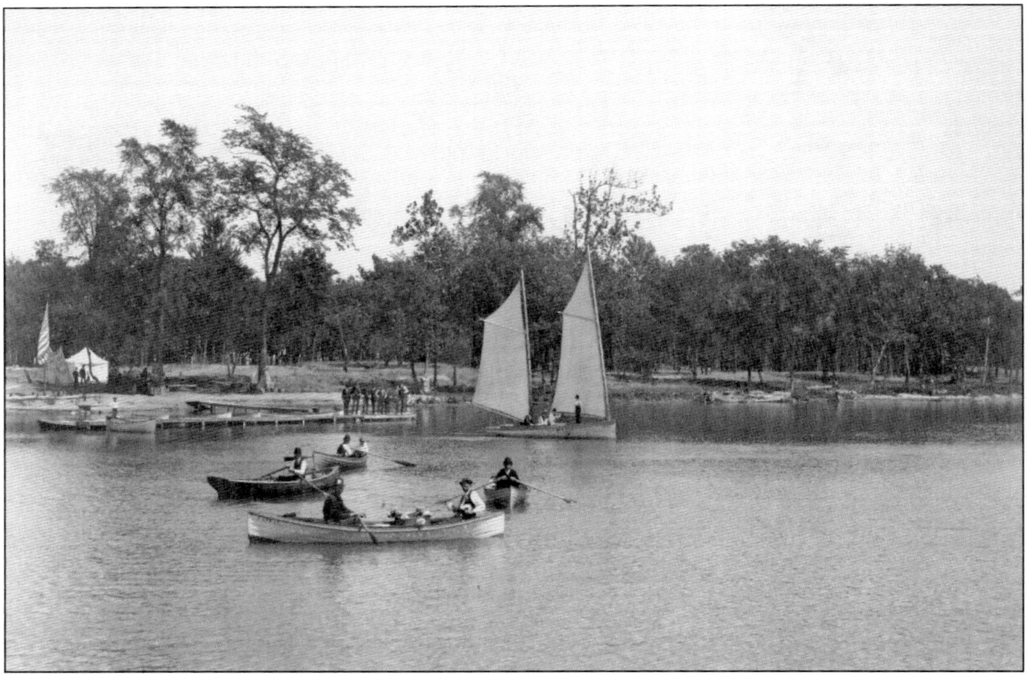

The Sandusky Bay side of Cedar Point offered a sheltered area, ideal for small sailboats and rowboats. Boating was a popular activity for visitors to the peninsula. Note the American flag in front of the tent. The peninsula was heavily wooded at the time of this c. 1885 photograph. (Courtesy of HPC.)

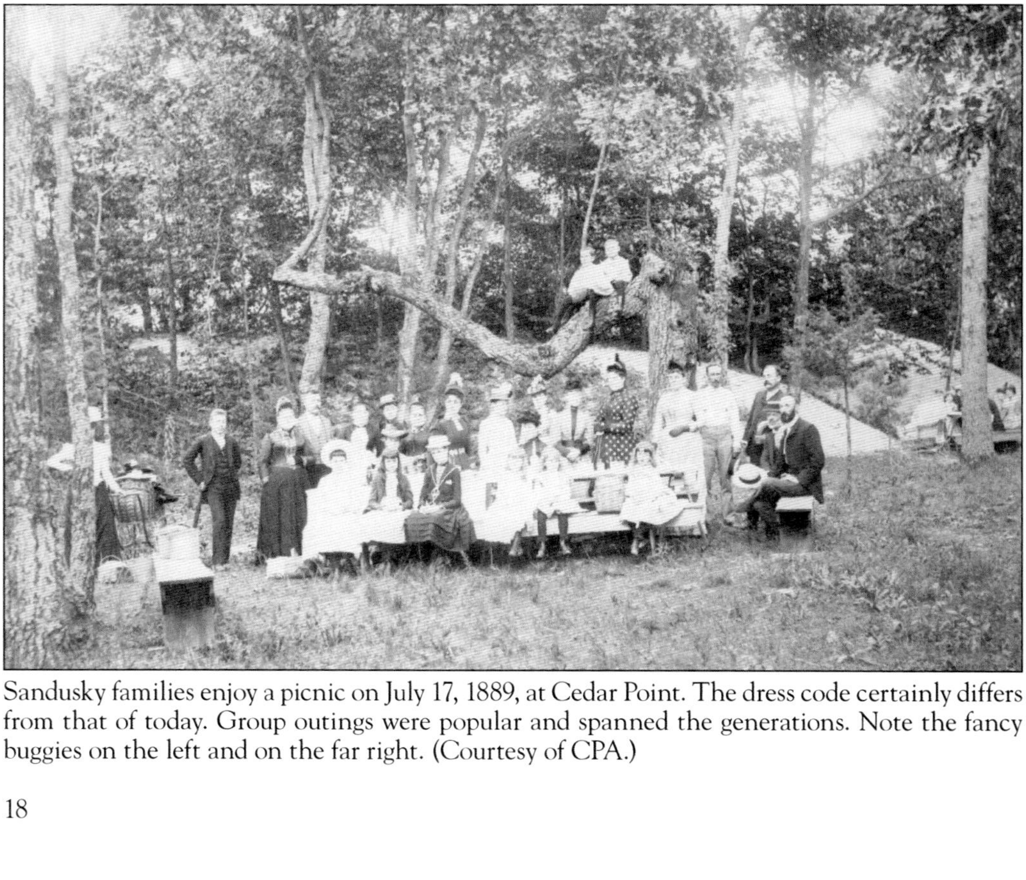

Sandusky families enjoy a picnic on July 17, 1889, at Cedar Point. The dress code certainly differs from that of today. Group outings were popular and spanned the generations. Note the fancy buggies on the left and on the far right. (Courtesy of CPA.)

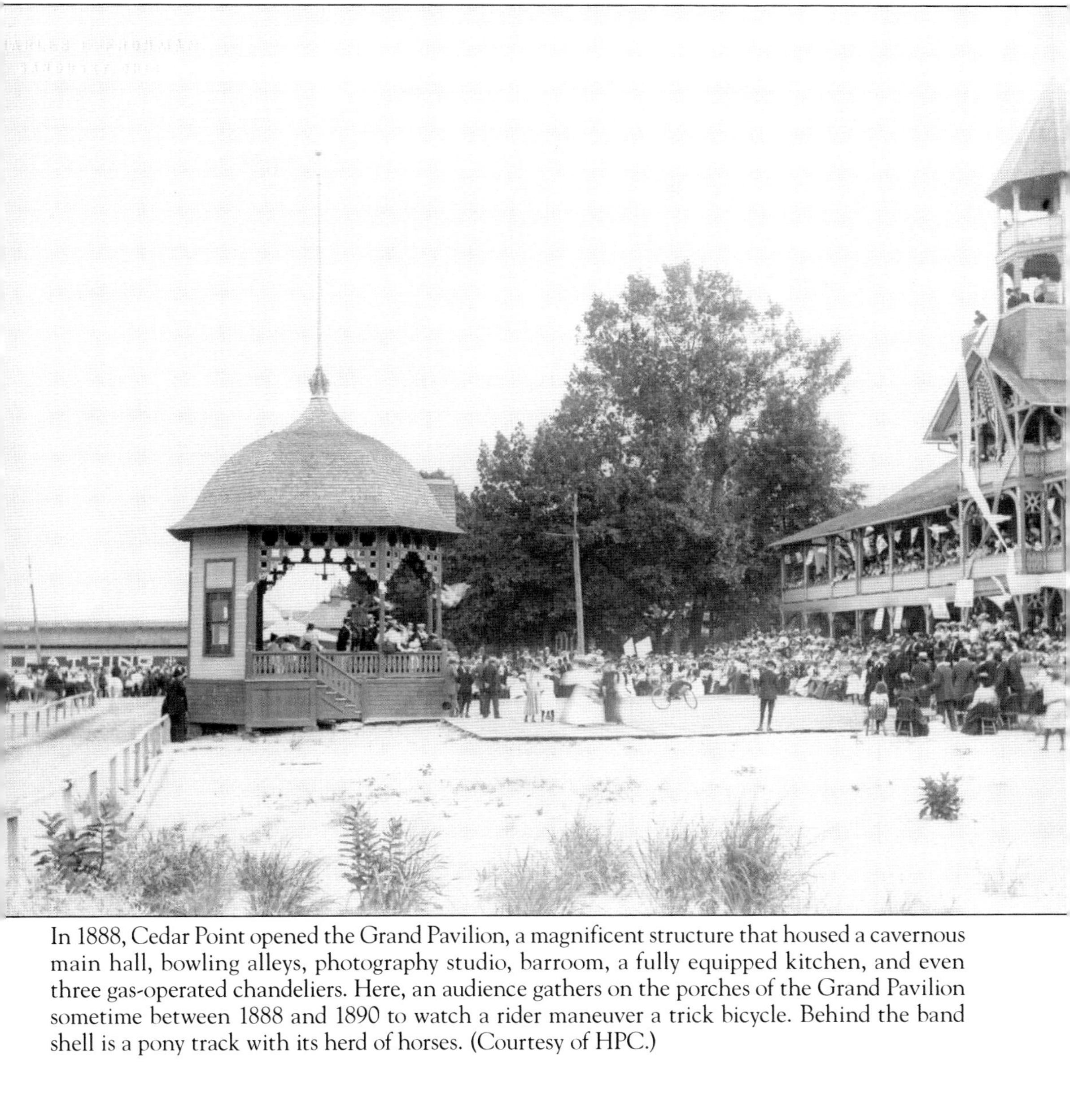

In 1888, Cedar Point opened the Grand Pavilion, a magnificent structure that housed a cavernous main hall, bowling alleys, photography studio, barroom, a fully equipped kitchen, and even three gas-operated chandeliers. Here, an audience gathers on the porches of the Grand Pavilion sometime between 1888 and 1890 to watch a rider maneuver a trick bicycle. Behind the band shell is a pony track with its herd of horses. (Courtesy of HPC.)

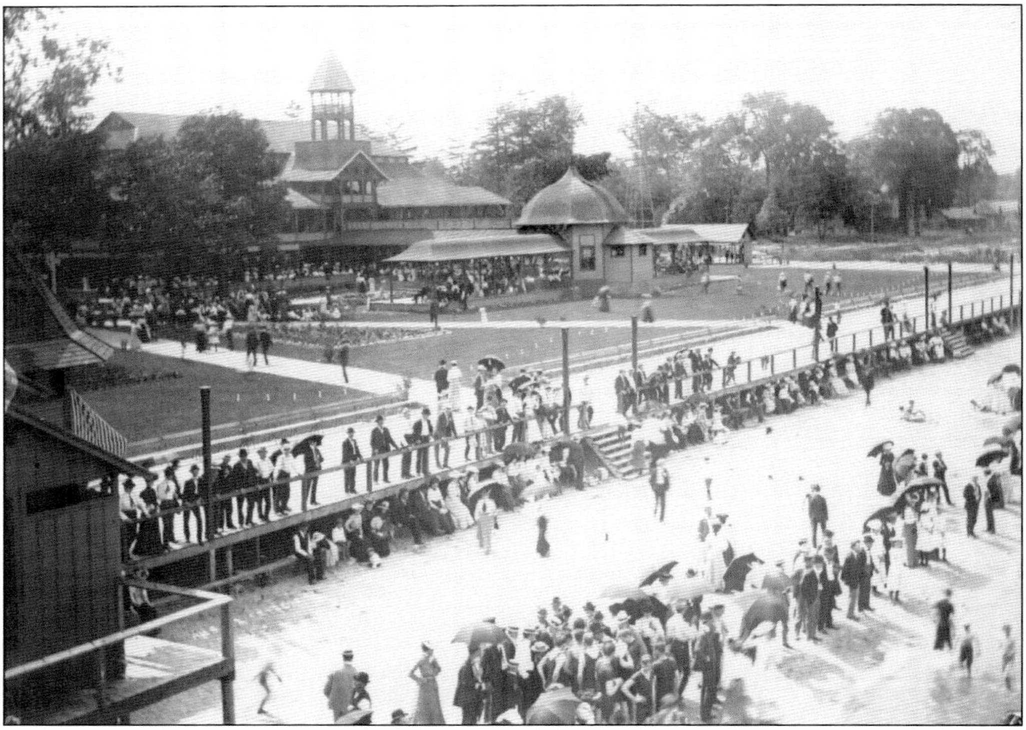

This c. 1890 photograph of the Grand Pavilion, taken from the lakeside, shows the band shell and boardwalk. The Grand Pavilion was the center of activity on the peninsula for many seasons. Beautiful floral displays were located on the grounds. (Courtesy of HPC.)

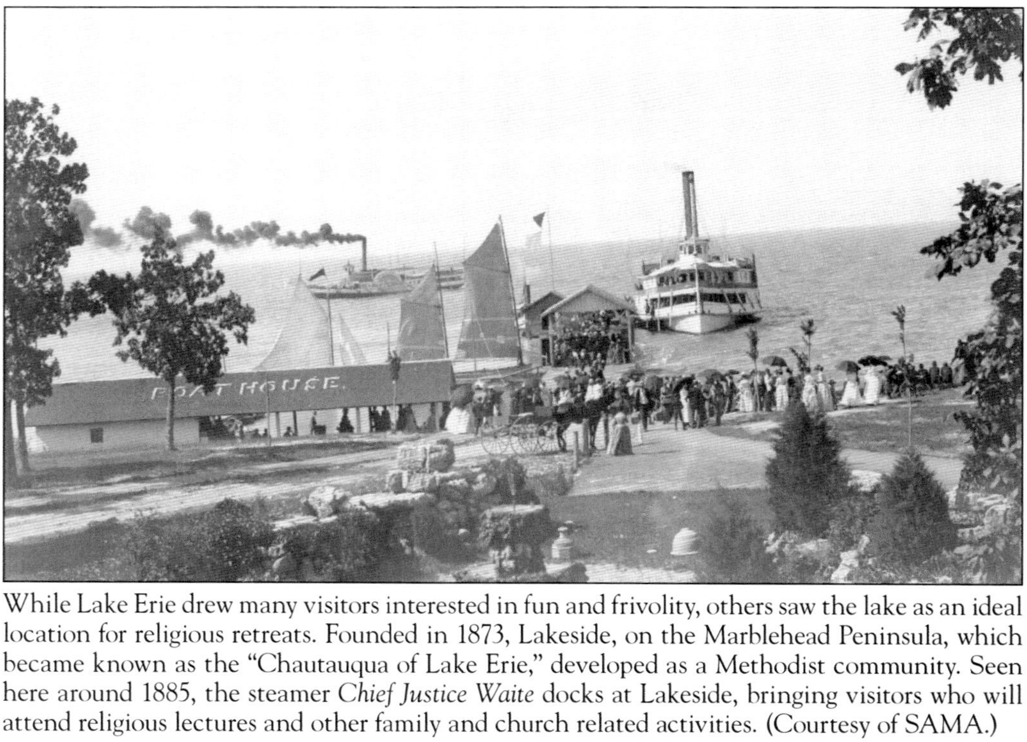

While Lake Erie drew many visitors interested in fun and frivolity, others saw the lake as an ideal location for religious retreats. Founded in 1873, Lakeside, on the Marblehead Peninsula, which became known as the "Chautauqua of Lake Erie," developed as a Methodist community. Seen here around 1885, the steamer *Chief Justice Waite* docks at Lakeside, bringing visitors who will attend religious lectures and other family and church related activities. (Courtesy of SAMA.)

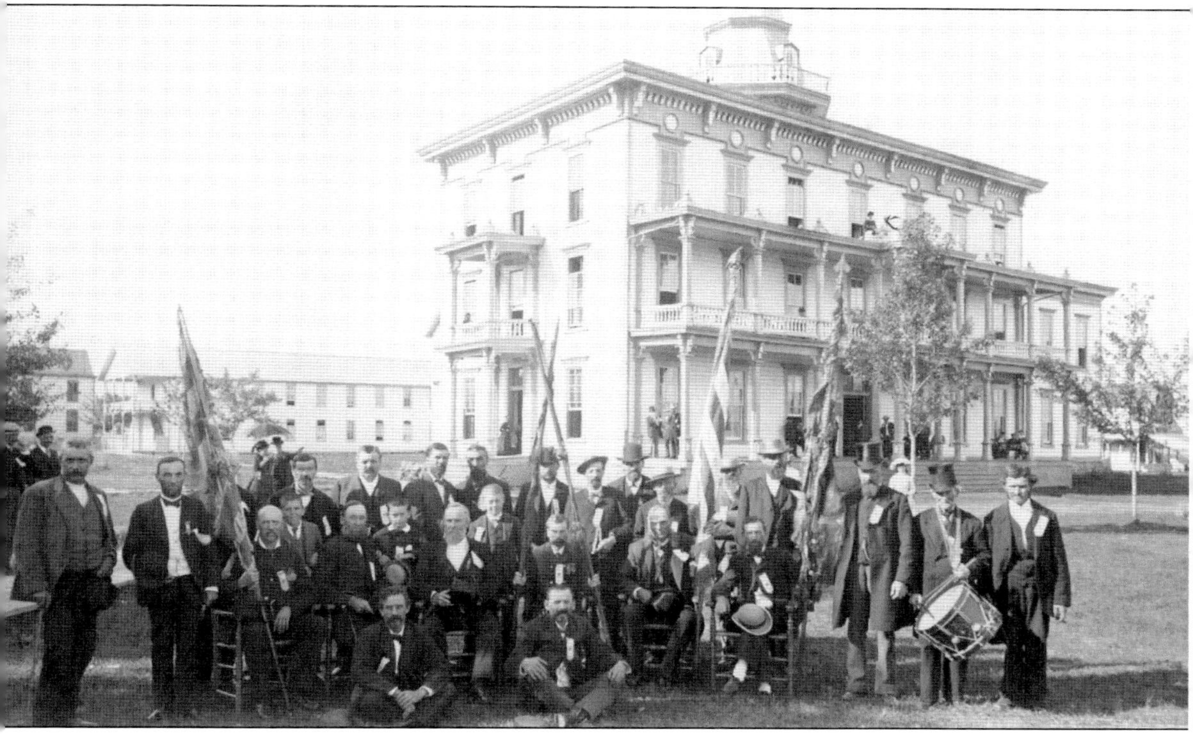

Opened in 1875, the Hotel Lakeside was a popular spot for conventions, reunions, and other events. This 1883 photograph shows a reunion of the 23rd Ohio Volunteer Infantry. Pres. Rutherford B. Hayes, the former commander of the regiment, is seated in the second row, left of center, wearing a starched white collar. Hayes reportedly favored Lakeside for these reunions because of its strict no alcohol policy. (Courtesy of HPC.)

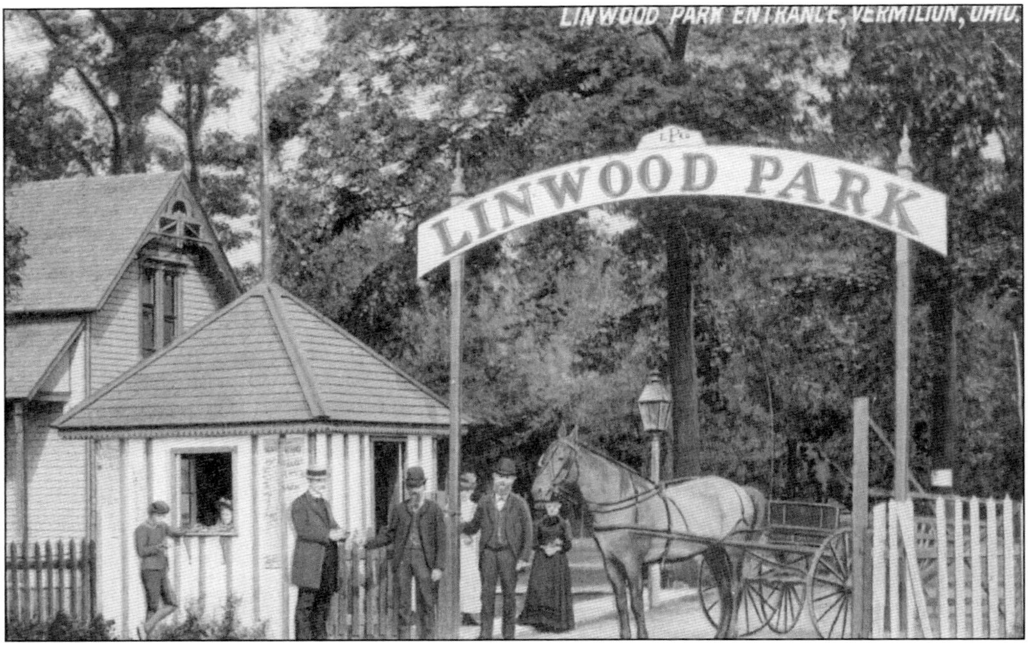

The brainchild of Vermilion resident and shoemaker Nicholas Wagner, Linwood Park, on the east side of the Vermilion River, opened in 1880. In addition to a beautiful beach, Linwood offered a dance hall, refreshment stand, rifle range, and pier. Wagner was likely inspired by the success of nearby Cedar Point. In 1883, he sold the property to a religious group headed by Bishop William Horn of Cleveland as a location for religious retreats. (Courtesy of Vermilion Archival Society.)

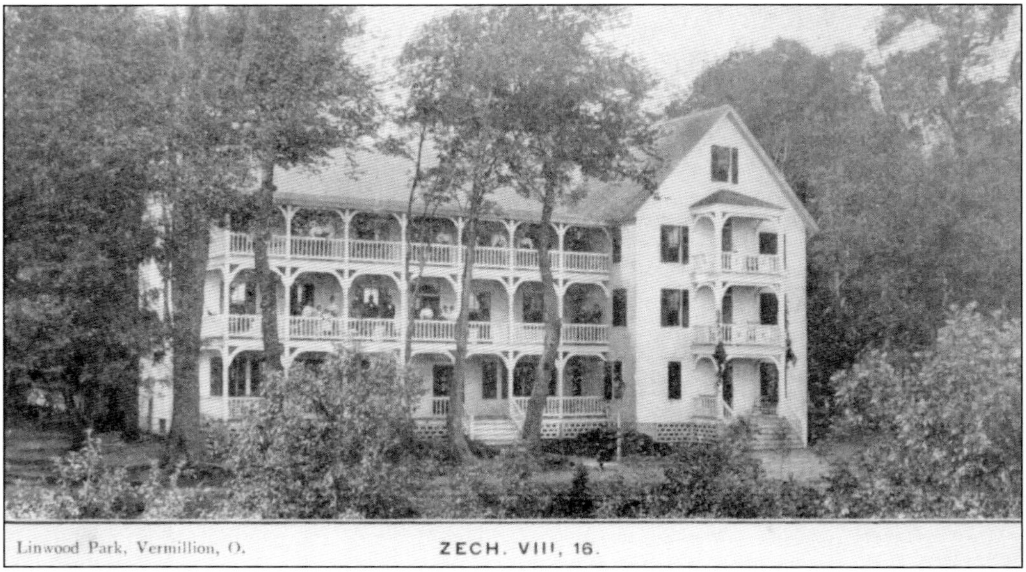

The Hotel Linwood, built in 1886, was a three-story, wood-frame building that stood until 1965. The hotel, seen here around 1885, was open only during the summer months and was very popular with visitors to the park. In this era, hotel stays tended to be longer, often two weeks or more. (Courtesy of Vermilion Archival Society.)

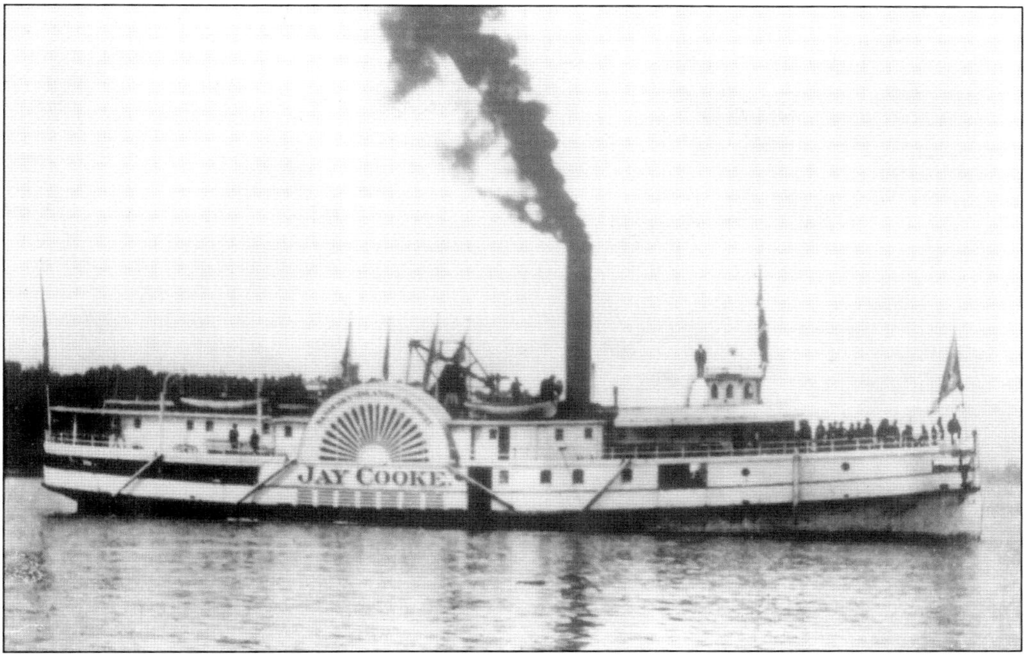

Named for the Sanduskian who sold bonds to finance the Civil War, the steamer *Jay Cooke* was built in Detroit in 1868 by John P. Clark. The vessel, seen here around 1880, ran between Detroit and Sandusky, with stops at Put-in-Bay and Cedar Point. In 1880, the ship was sold to the Sandusky & Islands Steamboat Company and underwent extensive repairs. In 1888, it was rebuilt and renamed the *City of Sandusky*. After only six more years of service, it was dismantled in 1894. In 1908, her hull was taken to Lake St. Clair for use as a break wall. (Courtesy of SLARC.)

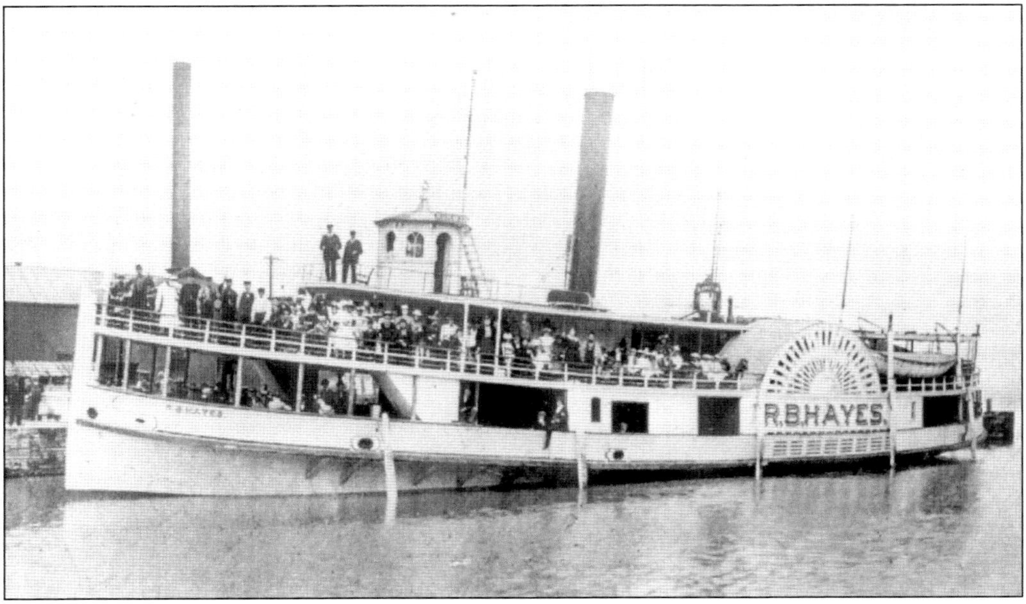

The steamer *R.B. Hayes* was built in 1876, the year Ohioan (and Fremont resident) Rutherford B. Hayes was elected president of the United States. Constructed by J.E. Monk of Sandusky, the ship provided passage to the Lake Erie islands, Cedar Point, and mainland ports until the 1920s. (Courtesy of HPC.)

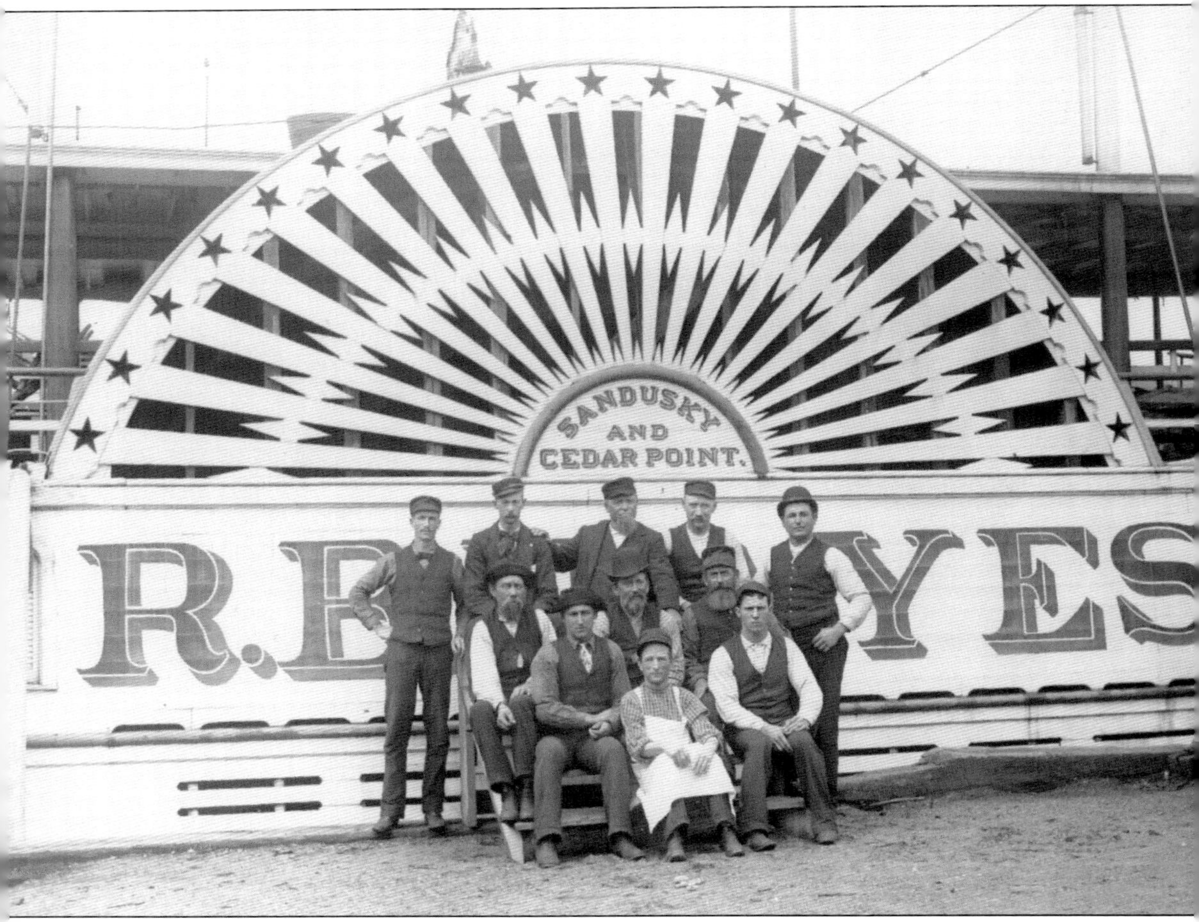

Crew members of the *R.B. Hayes* pose in front of the steamer around 1885. The ship's cook is easily identified by his white apron. Lake steamers needed large crews and as a group employed hundreds of men during the summer months. The *R.B. Hayes* was in operation for nearly half a century, a long life for a lake steamer. (Courtesy of SLARC.)

Two

Lake Erie's First Golden Age 1891–1910

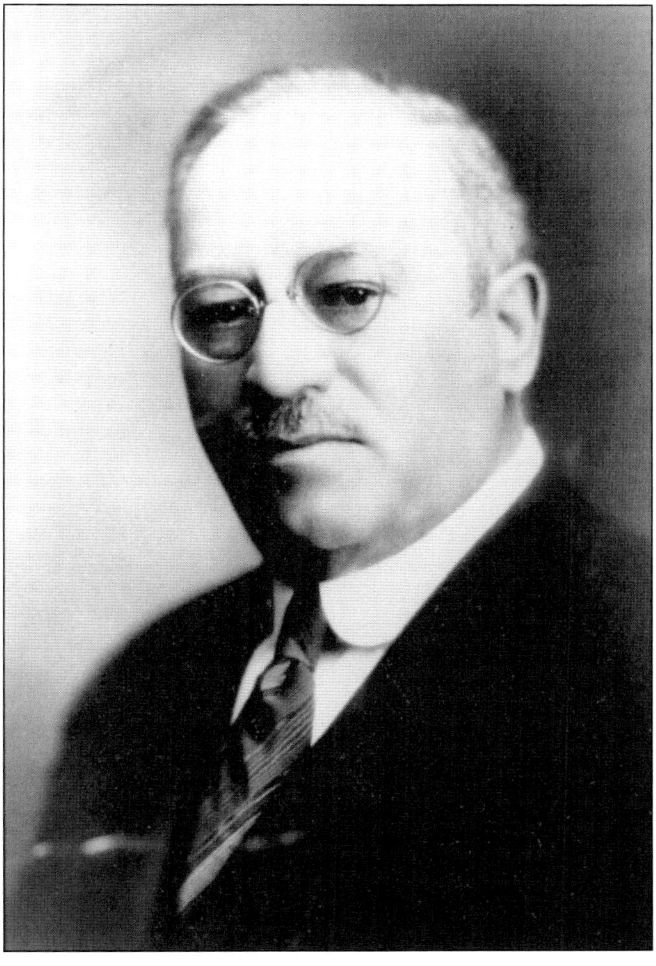

Entrepreneur, promoter, showman, executive, dreamer, George A. Boeckling, a native of Michigan City, Indiana, was the man who led Cedar Point into its first golden age. Gaining control of Cedar Point in 1897, he immediately embarked on a building program that within a decade made Cedar Point a regional, even national, destination. Under Boeckling's direction, Cedar Point added the massive Hotel Breakers in 1905, the Coliseum in 1906, and an electric-powered midway. The site had become both a resort and an amusement park. Boeckling ran Cedar Point until his death in 1931. (Courtesy of CPA.)

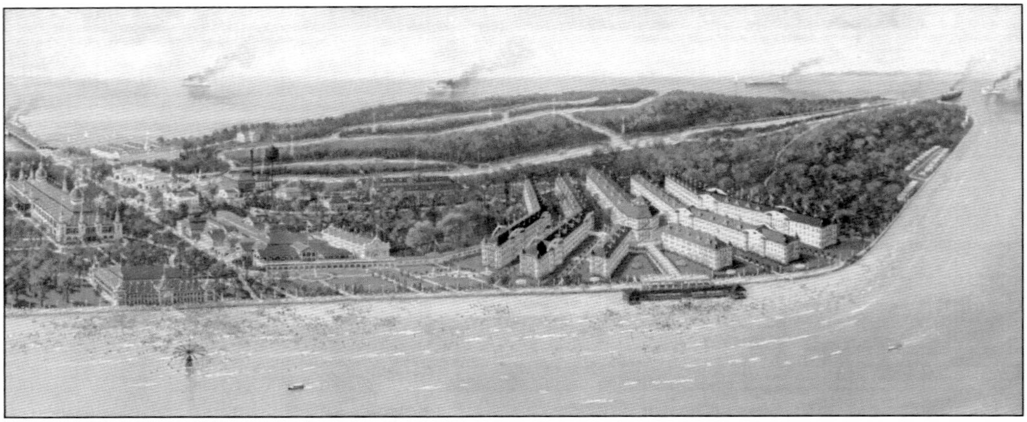

This idealized bird's-eye view of Cedar Point, produced around 1910, depicts a resort destination eager to impress. This image was used extensively in advertising, brochures, and other materials to promote the fast-growing resort. (Courtesy of Ohio History Connection.)

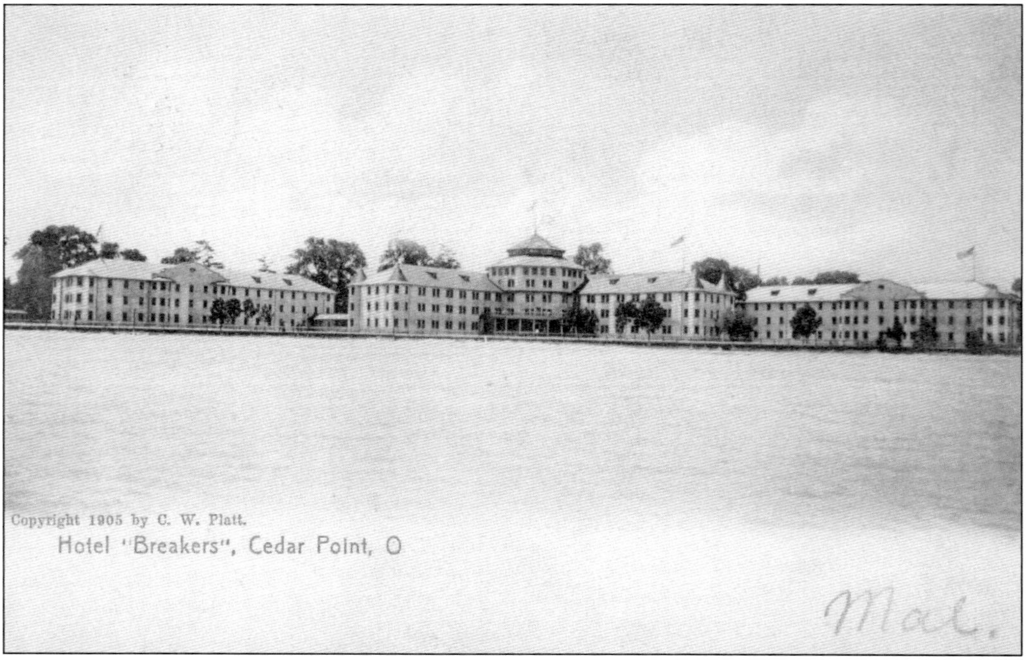

Hotel Breakers, pictured here in a C.W. Platt 1905 postcard, was George Boeckling's most impressive and favorite creation. Many of the 600 rooms offered stunning lake views. The hotel's public areas included a lobby with Tiffany chandeliers reputedly designed by the famous glass cutter Louis Buser. (Courtesy of Frank Landl.)

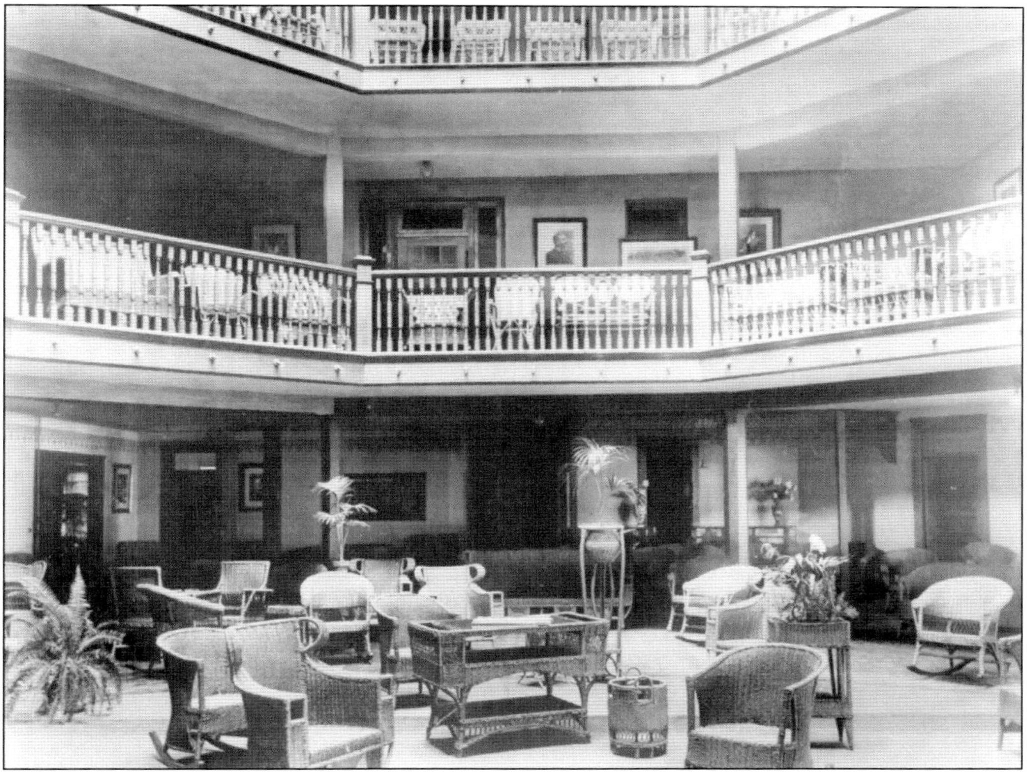

The five-story Rotunda, located off the lobby of the Breakers, provided guests with a special place to relax and enjoy the comforts of the hotel, including wicker furniture imported from Europe. Boeckling used the Rotunda for special entertainment, including impromptu concerts from Metropolitan Opera stars like Enrico Caruso and Nellie Melba, who visited Cedar Point on their way to Chicago for the Met's summer season. (Courtesy of HPC.)

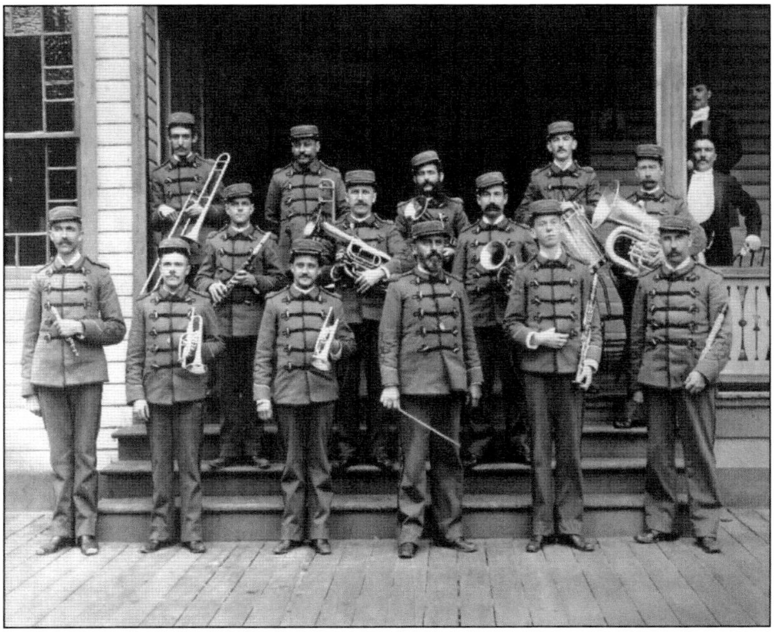

Headed by Charles Baetz, the Sandusky-based, 14-member Great Western Band was one of Cedar Point's main musical attractions in the 1880s and 1890s. This photograph was probably taken in front of the Grand Pavilion. The two men in white tie and tails at right are most likely restaurant managers. (Courtesy of SLARC.)

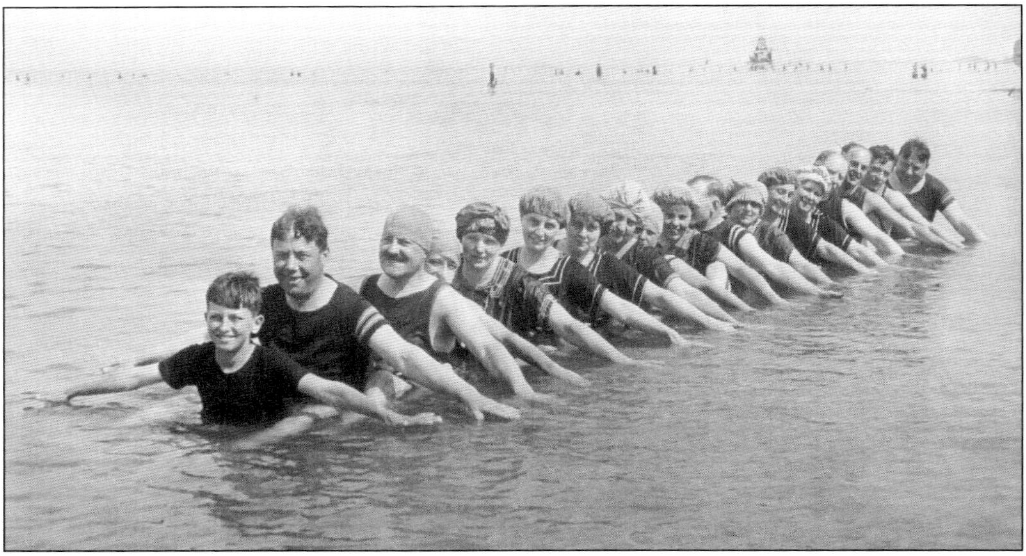

Cedar Point's beach was a primary attraction from the earliest days of the resort. Victorian and Gilded Age swimwear certainly differed from today's styles, especially the popularity of bathing caps. But, swimmers still managed to have fun, as seen here around 1905. Note the water swing in the background. (Courtesy of CPA.)

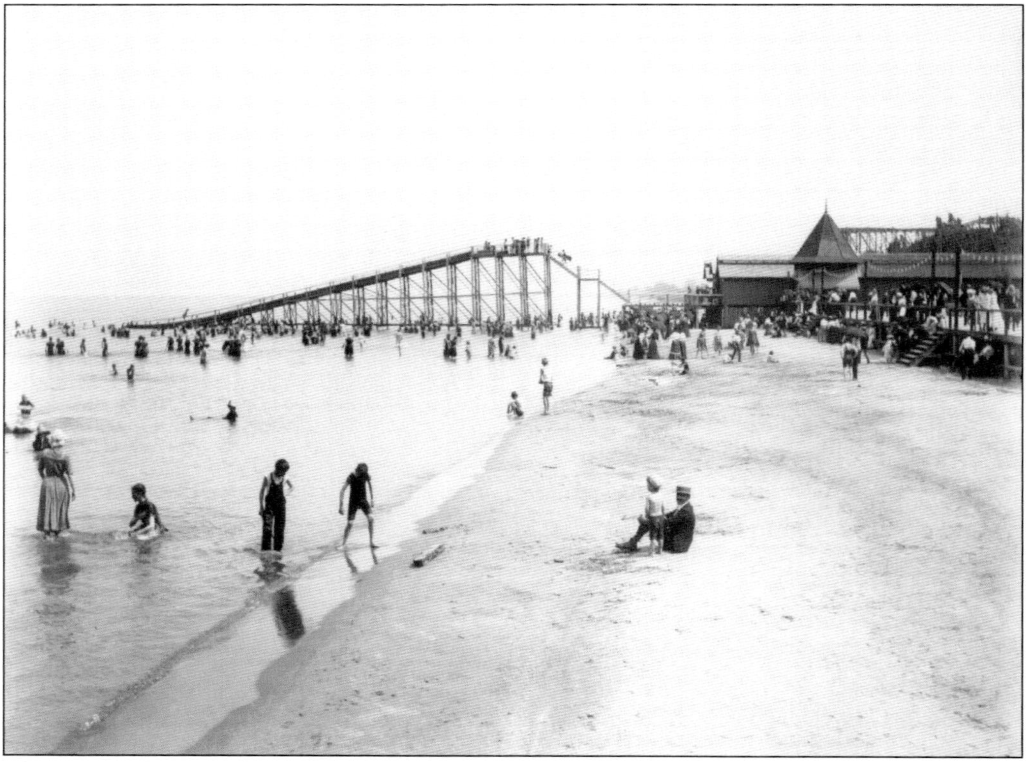

This c. 1895 photograph of Cedar Point Beach includes the latest in water slides in the background. Sea slides, also called water toboggans, were very popular on beaches in the shores and islands region in the 1890s and in the first two decades of the 20th century. Built mainly of wood, they required extensive maintenance to withstand the rigors of Lake Erie winters. (Courtesy of HPC.)

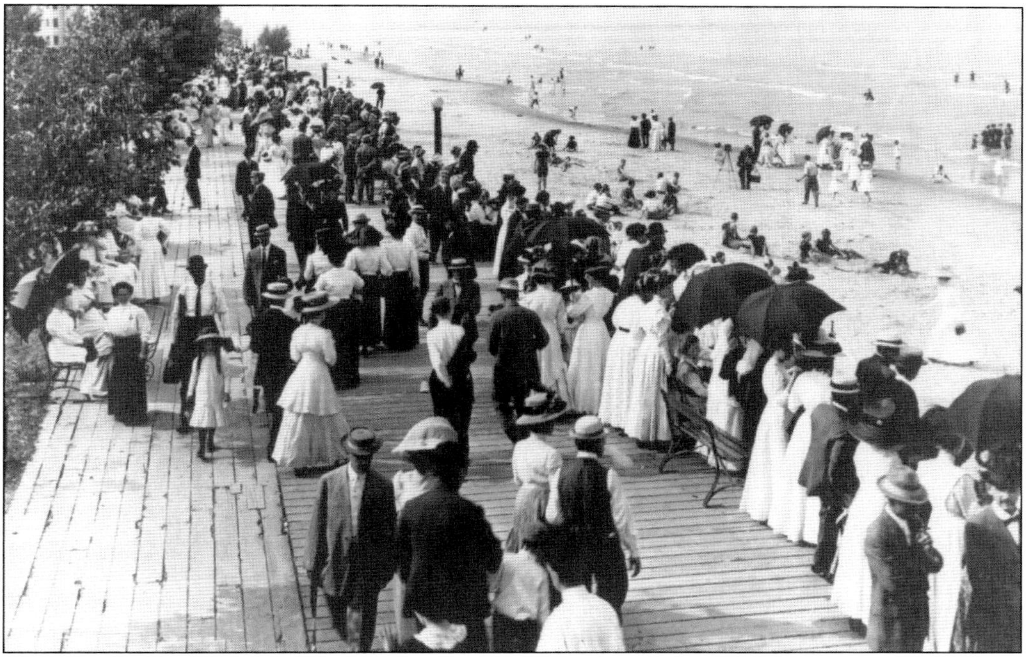

Guests fill the boardwalk and beach in front of the Hotel Breakers on a beautiful summer day sometime between 1900 and 1910. A photographer has set up equipment on the beach. Interestingly, there are many more guests on the boardwalk than on the beach. "Strolling the boardwalk" was a major activity at Cedar Point. (Courtesy of HPC.)

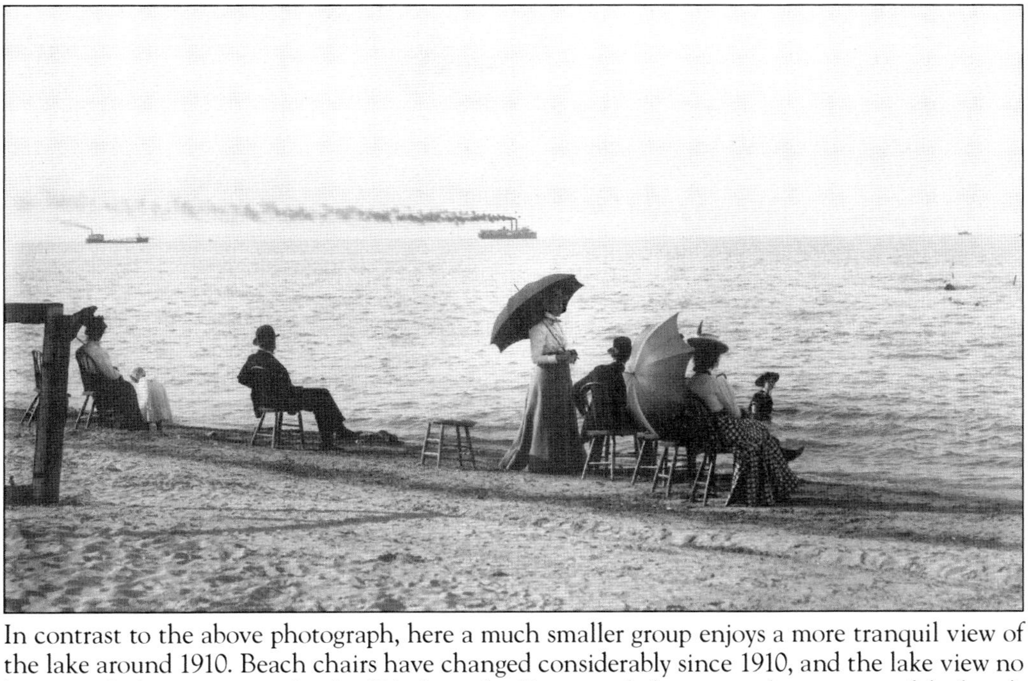

In contrast to the above photograph, here a much smaller group enjoys a more tranquil view of the lake around 1910. Beach chairs have changed considerably since 1910, and the lake view no longer includes streaming clouds of black smoke. For many ladies, parasols were part of the beach uniform in the early 1900s. (Courtesy of SAMA.)

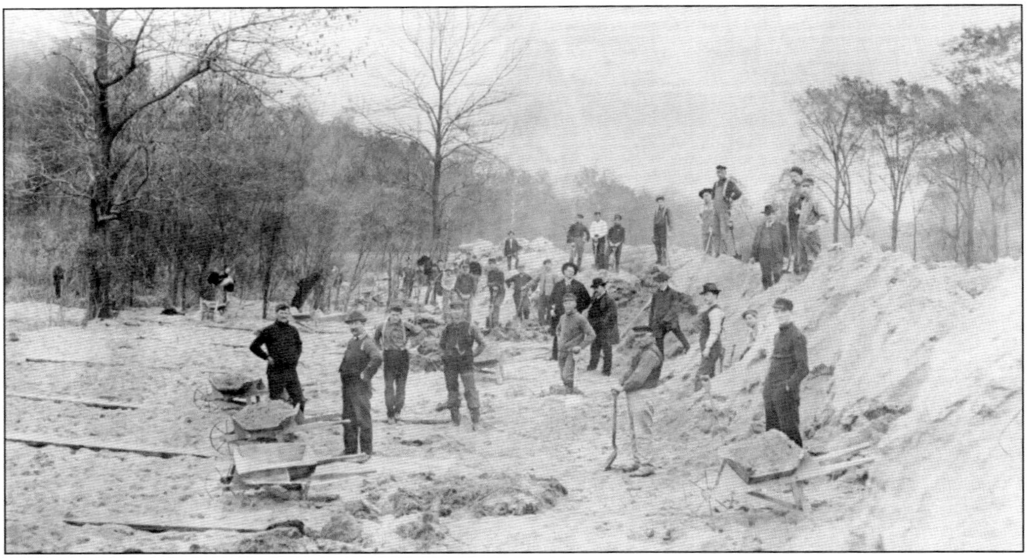

At the close of the 1904 season, George Boeckling signed a $25,000 contract with the Detroit Dredging Company to create a series of lagoons on the property. The lagoon network served several purposes. Resort guests could use water transportation to get from the boat docks to the Hotel Breakers. The lagoons also allowed easier transport of coal to the new electric power station. Finally, the lagoons were a great amenity for guests to enjoy on their own. Boeckling personally supervised the construction project, specifying that the depth be at least 10 feet. Much of the work was done by hand, as seen in this rare photograph. (Courtesy of HPC.)

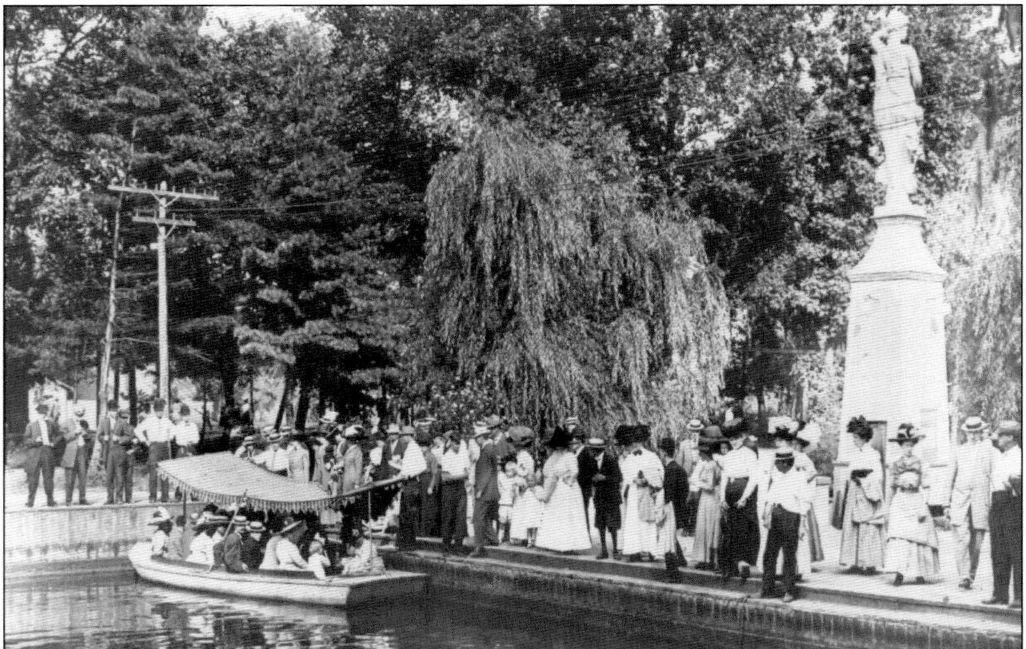

Guests embark on a boat trip in the Cedar Point lagoons around 1910. The statue at right was one of a number of statues Boeckling purchased at the 1904 St. Louis World's Fair and had installed on the Cedar Point grounds. Some are still on display in the park. Electricity had come to Cedar Point, as evidenced by the transmission pole in the left background. (Courtesy of HPC.)

With the new Cedar Point Bathhouse in the background, these three women stroll the midway, enjoying their visit to the resort. The facade atop the building advertises the latest styles in swimwear for ladies. This photograph was taken sometime between 1910 and 1912. (Courtesy of HPC.)

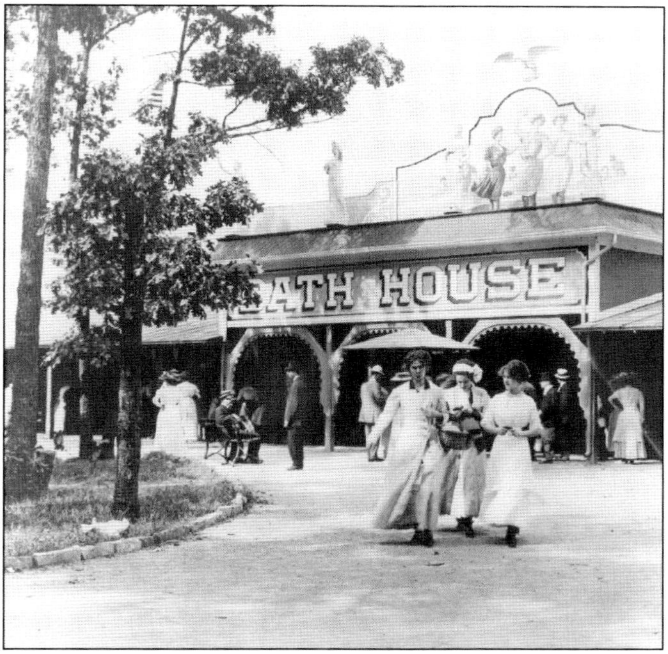

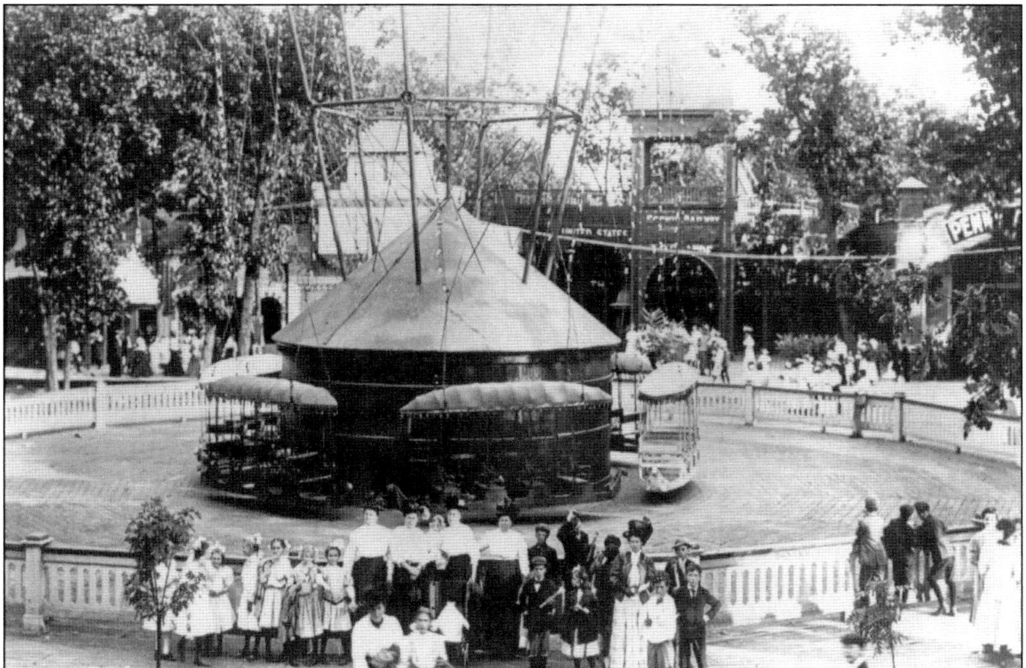

The centerpiece of Cedar Point's Amusement Circle was the Circle Swing, manufactured by the Ingersoll Company. The ride was very popular with resort guests. In this c. 1907 photograph, the group of guests in the foreground is unknown, but it comprises mostly mothers and children. Note the "Danger" message posted on the fence and the wooden plank ground cover surrounding the ride. George Boeckling understood the appeal of the beach, the picnic grounds, and the hotels, but he also knew that for Cedar Point to grow and prosper, midway rides and attractions had to be part of the offering. (Courtesy of HPC.)

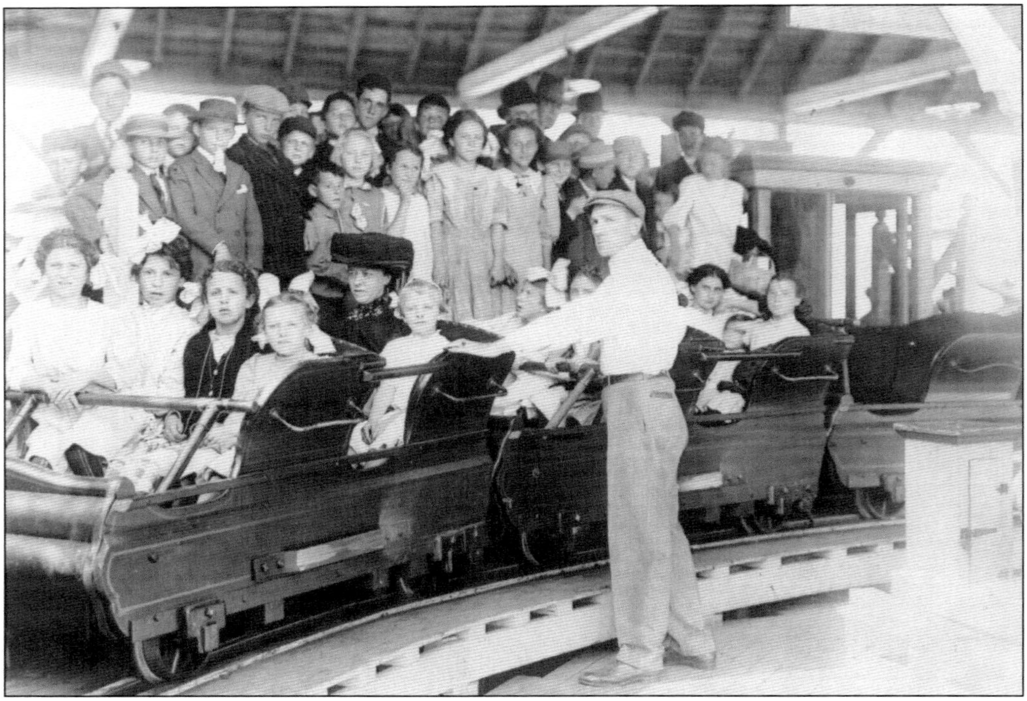

The Racer, erected along the beach in 1902, was George Boeckling's first roller coaster. Built by the Ingersoll Company of Pittsburgh, the Racer reached a rather modest height—certainly by today's standards—of 46 feet. A ride on the Racer cost a nickel. In this c. 1906 photograph, the female rider in the second row apparently is not concerned about losing her hat. The faces of the riders, with one exception, are fixated on the camera. (Courtesy of HPC.)

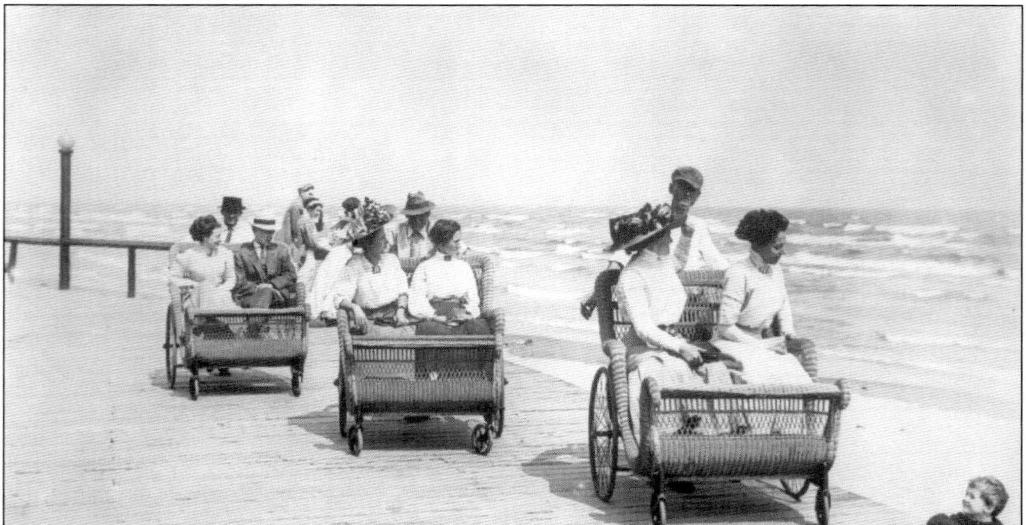

Roller cars, which first appeared on the Atlantic City Boardwalk, were popular at Cedar Point in the first decade of the new century. Visitors rented the wicker chair, and a guide pushed the chair along the boardwalk. Most patrons were women. Seen here around 1904, riders enjoy a view of Lake Erie as a young lad looks on. (Courtesy of HPC.)

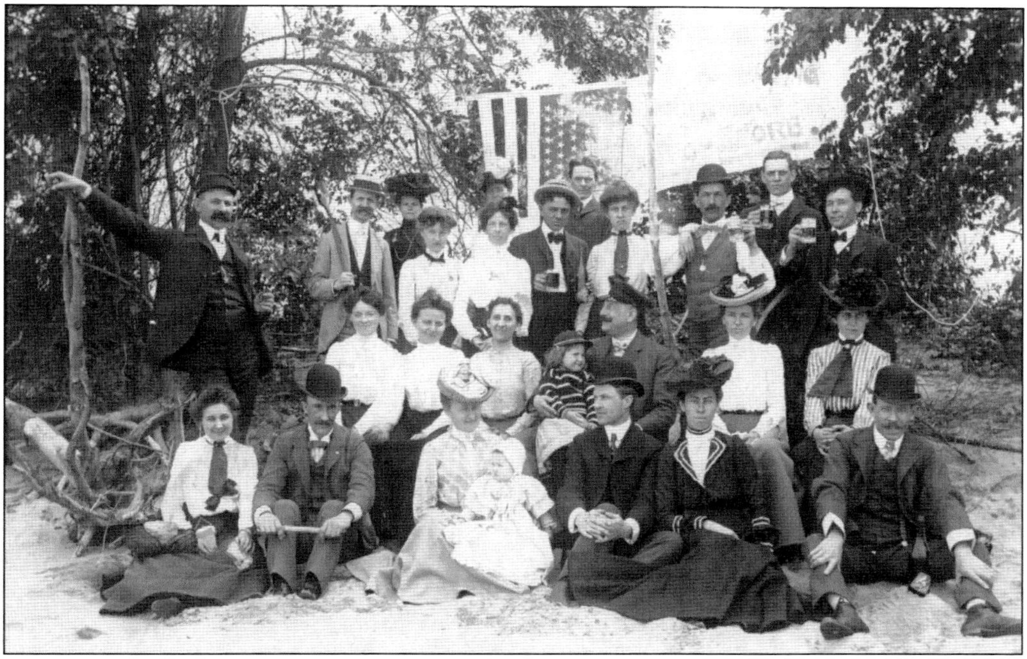

From its earliest days as a resort, Cedar Point was a popular location for company picnics and outings. Cedar Point's large groves of trees were ideal locations for picnics, and the park's management added tables, tents, and open-air structures to help attract groups. In this photograph, taken in 1902, the employees and families of the C.L. Engels Company pose during the company picnic. The store, which opened in Sandusky in 1887, was known as "Sandusky's Big Store." (Courtesy of SLARC.)

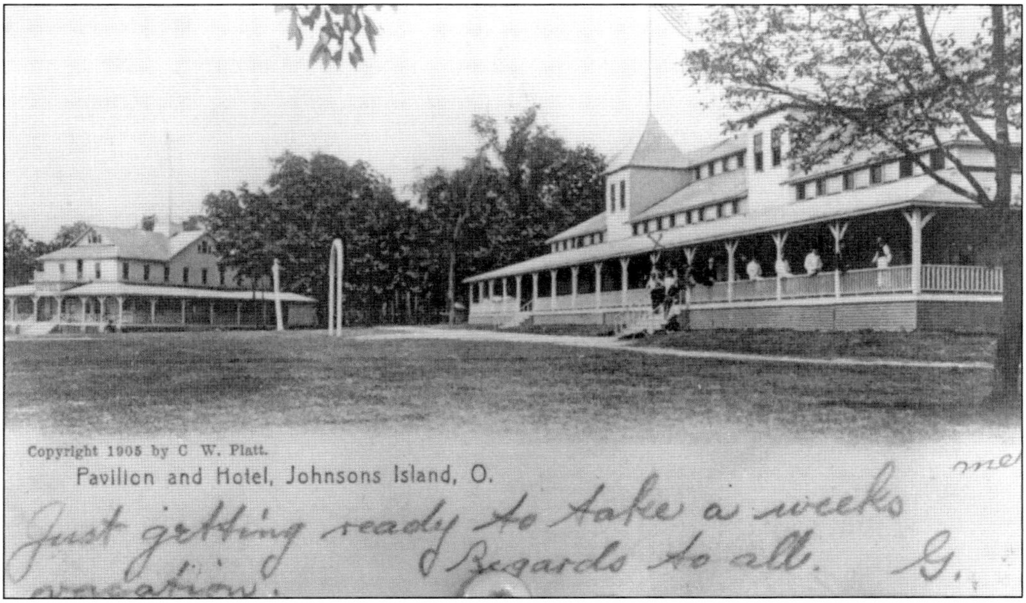

In its early days, Cedar Point's chief local competitor was the Johnson's Island Resort, which operated from 1894 until it was destroyed by fire in 1897. Though short-lived, the resort, seen here around 1895, impacted Cedar Point attendance with its newer facilities and upscale entertainment. The island is located in Sandusky Bay, a short distance from Cedar Point. The island was used as a prison for Confederate officers during the Civil War. (Courtesy of SLARC.)

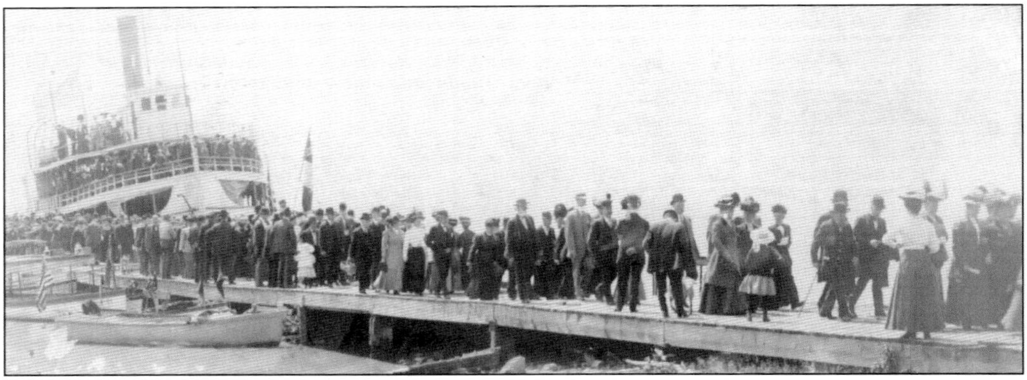

The steamer *R.B. Hayes* delivers passengers to the dock at Johnson's Island around 1895. The boat has a decided list to starboard as passengers try to get off as quickly as possible. Both ladies and gentlemen are dressed in their Sunday best for the day's events at the resort. (Courtesy of SAMA.)

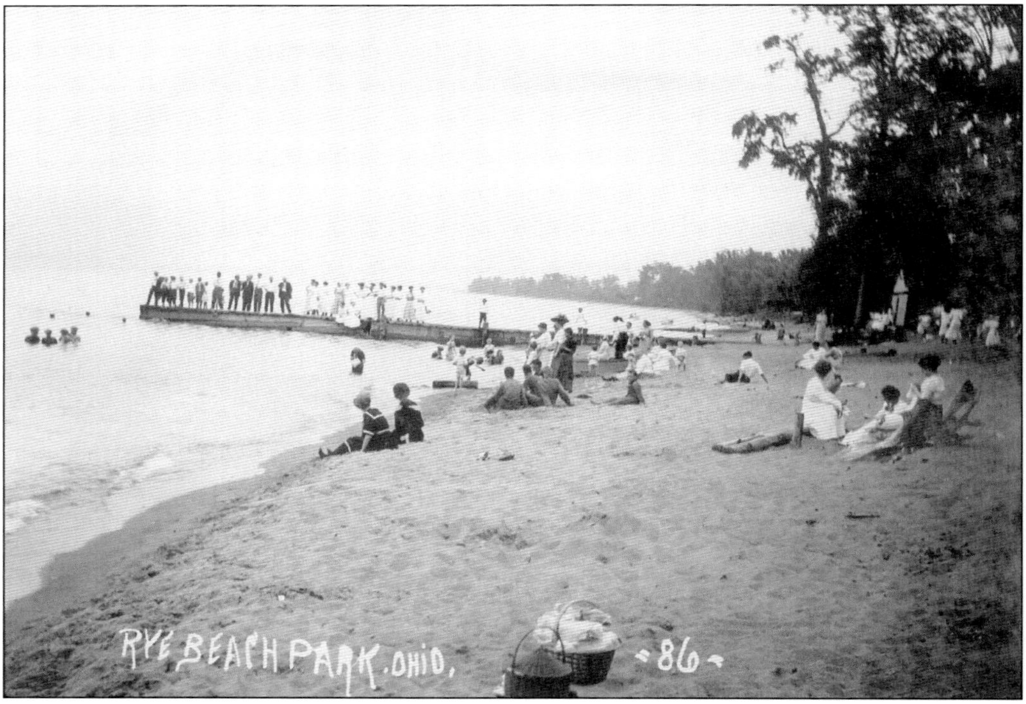

Rye Beach Park, which opened in 1893, was one of several cottage communities along Lake Erie between Sandusky and Vermilion. Located on the west side of Huron, Rye Beach was a popular summer getaway. By 1904, owner William Bruns had built more than 100 cottages. The park also featured a dance pavilion and a bowling alley. The beach is seen here around 1897. (Courtesy of HPC.)

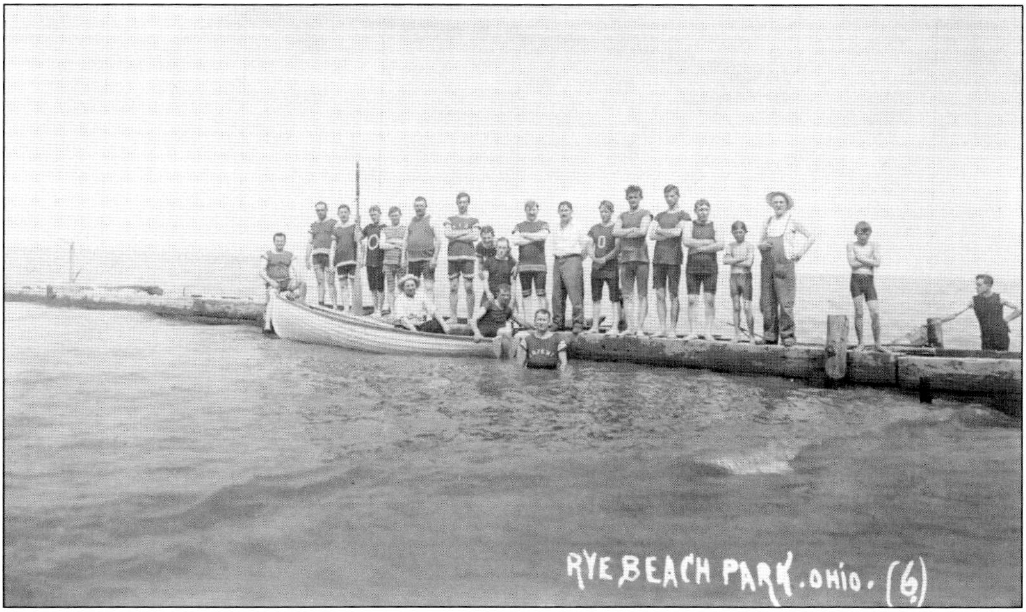

Rye Beach bathers pose for photographer Ernst Niebergall around 1900. For 20 years, Niebergall roamed the shores and islands region looking for subjects to photograph. He left behind a remarkable collection that documents the life of the region. His collection is housed at the Hayes Presidential Center in Fremont. (Courtesy of HPC.)

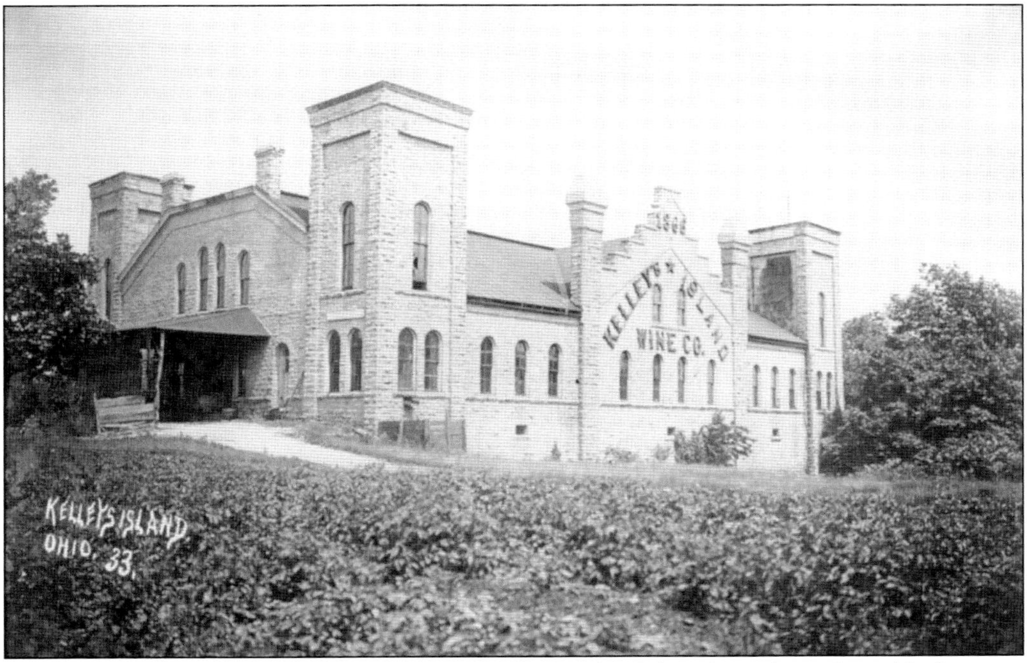

The climate on the Lake Erie shore and on the nearby islands has always been favorable for grape growing. Kelleys was the first island to develop a wine industry. The Kelley Island Wine Company, established in 1865, built its first cellar in 1872. The original winery burned down in 1876 and was replaced by an imposing limestone structure, seen here around 1900. It also succumbed to fire in 1915. (Courtesy of Leslie Korenko.)

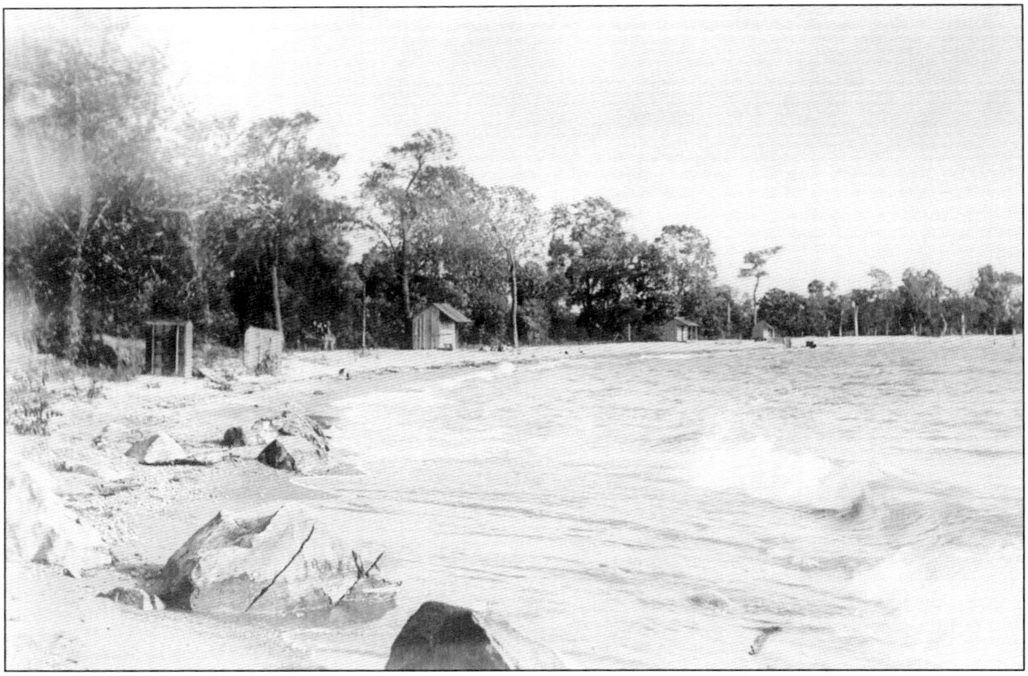

Several bathhouses dot the south shore of Kelleys Island in this 1894 photograph. Although rustic in appearance, these structures were functional, providing bathers a place to store clothing, picnic items, and other supplies. At the time, the south shore of the island was considered the best place for swimming. (Courtesy of HPC.)

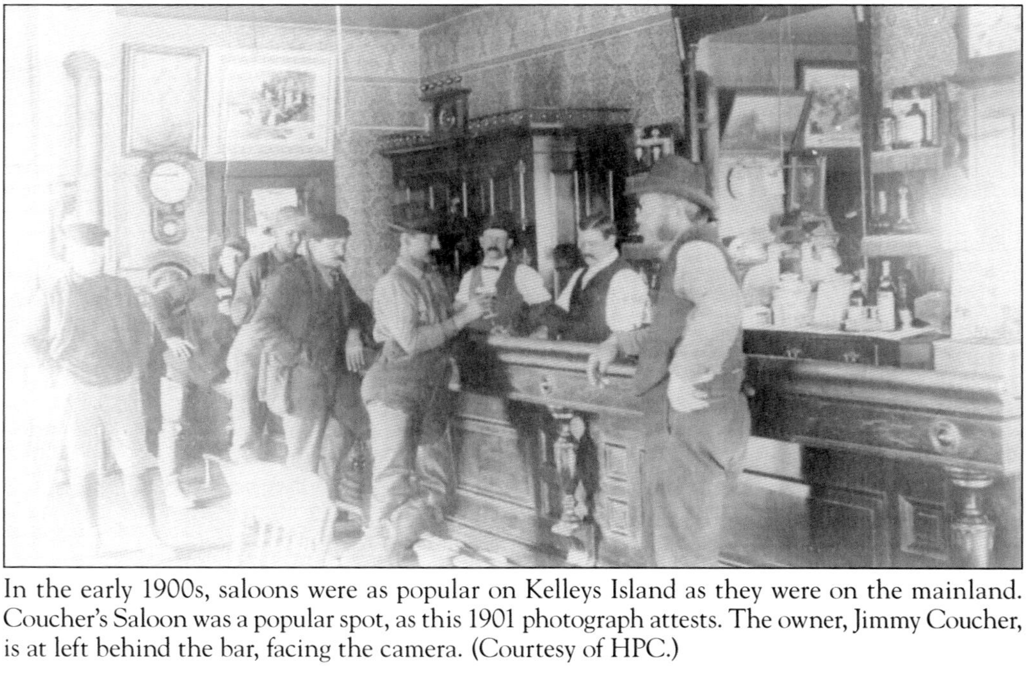

In the early 1900s, saloons were as popular on Kelleys Island as they were on the mainland. Coucher's Saloon was a popular spot, as this 1901 photograph attests. The owner, Jimmy Coucher, is at left behind the bar, facing the camera. (Courtesy of HPC.)

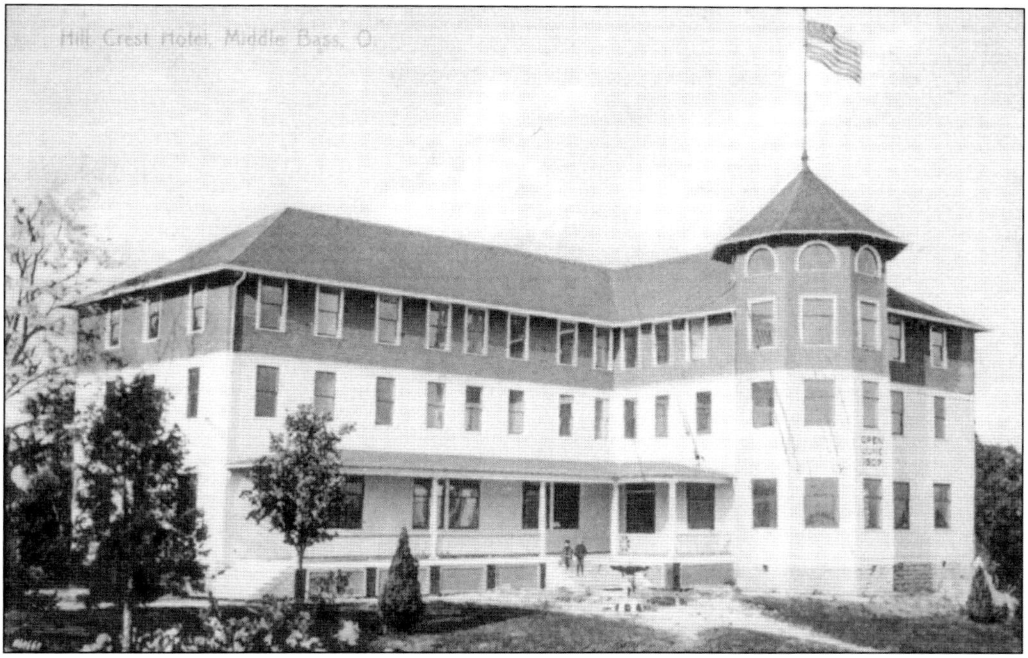

Andrew Wehrle built the 60-room Hill Crest Hotel in 1906 to accommodate the growing tourist trade to Middle Bass Island. Unfortunately, a fire in 1923 destroyed the hotel and the adjacent dance hall. Hotel fires were very common in the early 20th century, an era without smoke alarms and sprinkler systems and when buildings were made of wood. (Courtesy of SAMA.)

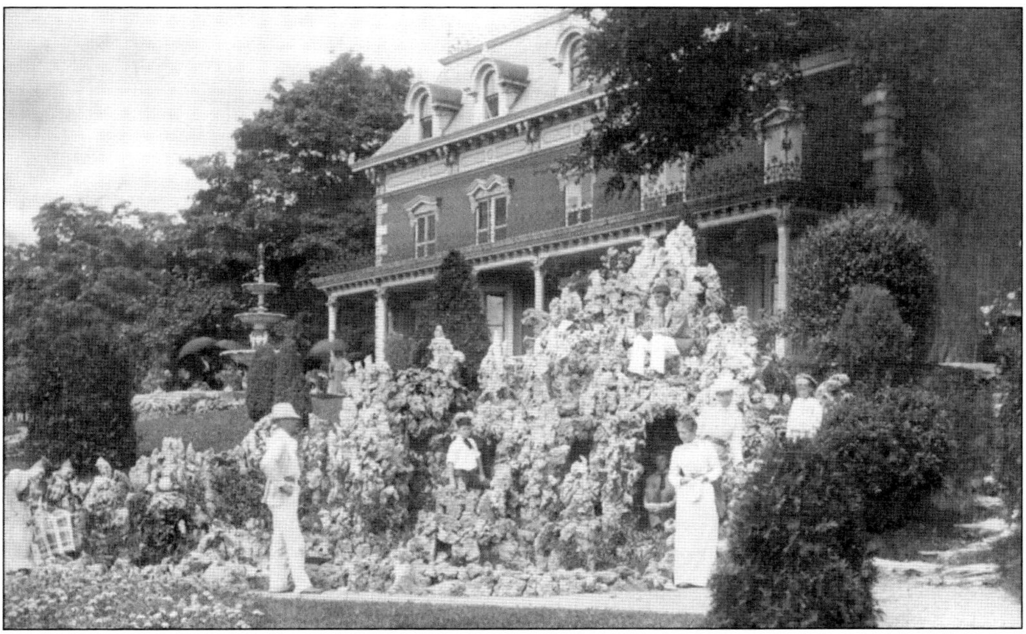

The Rockery was a unique rock formation in front of Andrew Wehrle's beautiful Middle Bass home. It was very popular with Middle Bass visitors. In the Cooke family journal, there are several references to visiting the Rockery at Wehrle's. Here, children pose at the limestone wall around 1892. The home burned down in 1905. Remnants of the Rockery are still visible outside the Lonz Winery building. (Courtesy of HPC.)

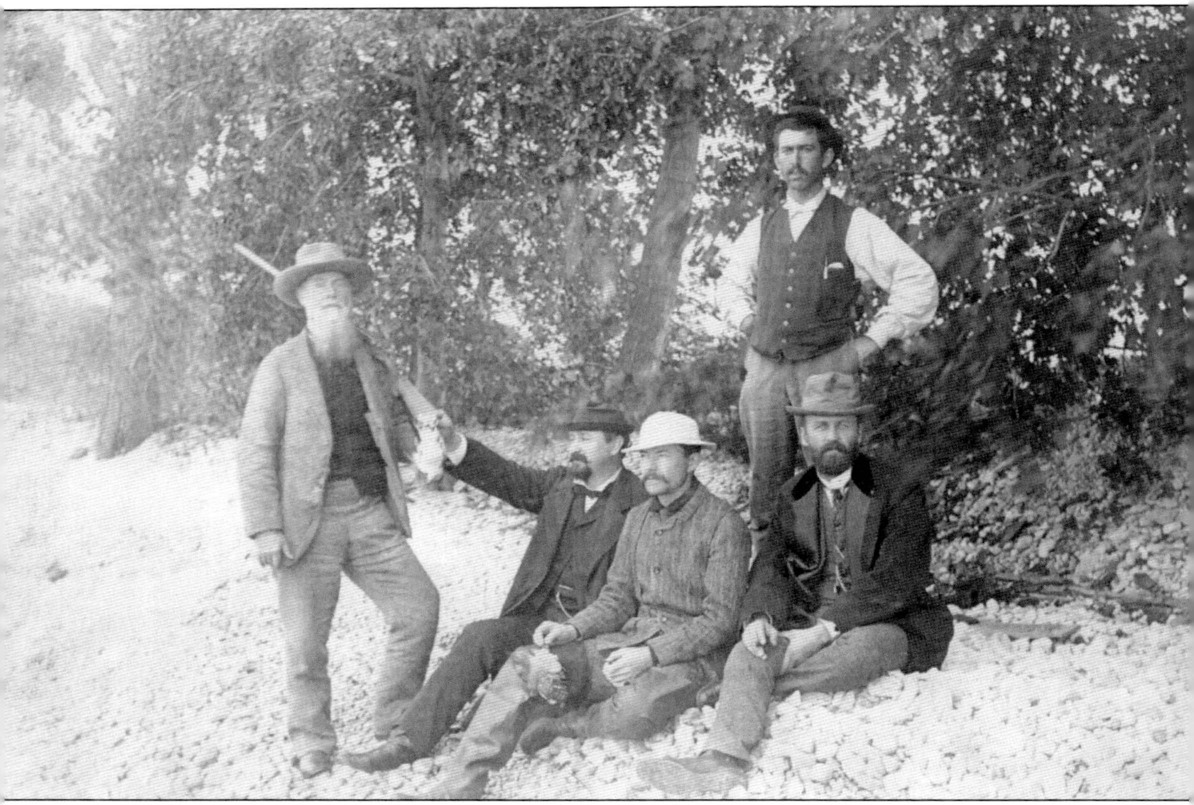

In addition to his fishing prowess, Jay Cooke (left) was an avid hunter, especially of ducks and other waterfowl. In the fall, the Lake Erie shoreline, Sandusky Bay, and the islands teem with migrating ducks. A number of hunting clubs in the area date from the late 19th century. This photograph of a hunting expedition was taken on Middle Bass around 1897. (Courtesy of HPC.)

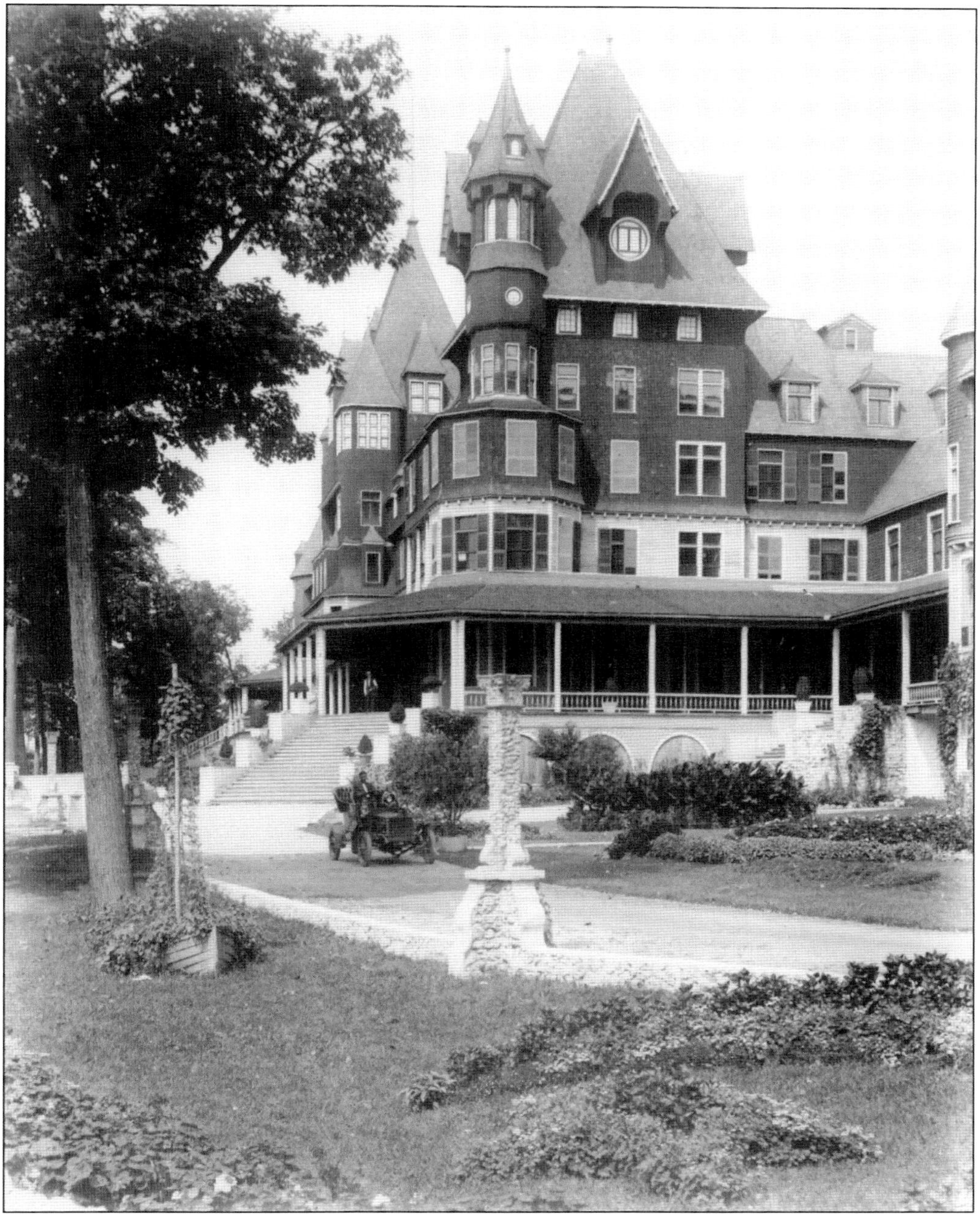

The palatial Hotel Victory opened on South Bass Island in 1892. An enormous structure, it could comfortably accommodate 625 guests and had capacity for 800 dinner guests. The owners believed the hotel, located on a bluff on the west side of the island, could compete favorably with the prestigious summer retreats on the East Coast. Promotional materials boasted that the Hotel Victory was the largest seasonal hotel in the country. Guests arrived from the Village of Put-in-Bay by electric tram. The hotel was the pinnacle of resort development on the Lake Erie islands, but it was not a financial success. The hotel burned to the ground in a massive fire in 1919. Today, the grounds are part of South Bass Island State Park. (Courtesy of David W. Francis collection.)

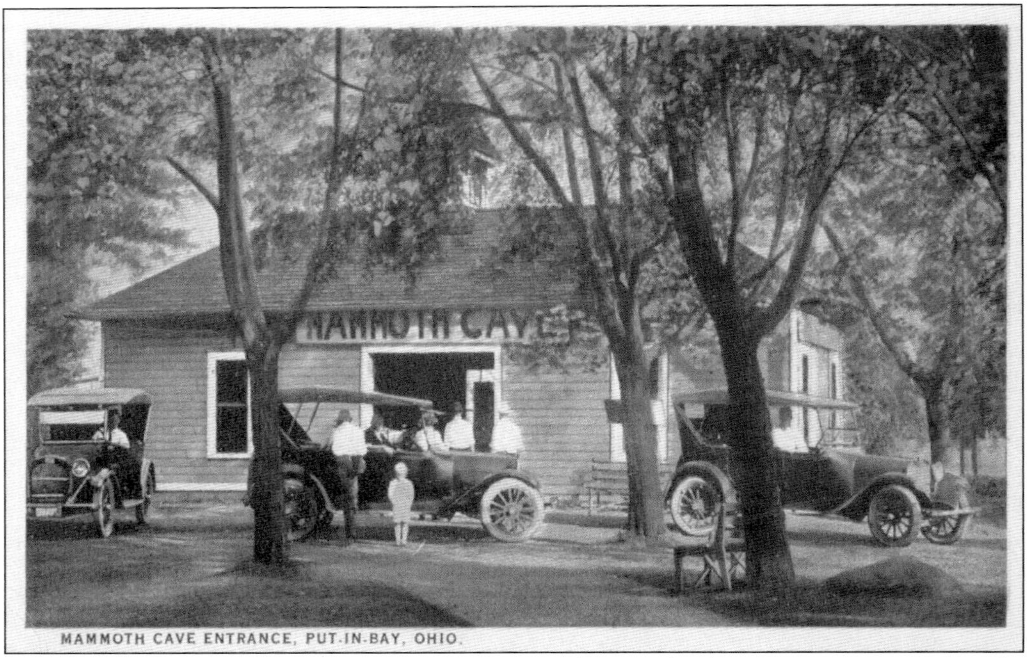

Discovered in 1886, Mammoth Cave was one of South Bass Island's main tourist attractions for decades until it closed to the public in 1955. The entrance led 60 feet underground to an 8-foot-by-60-foot lake. Due to their limestone base, the Bass islands have numerous caves. This postcard photograph was taken between 1900 and 1910. (Courtesy of HPC.)

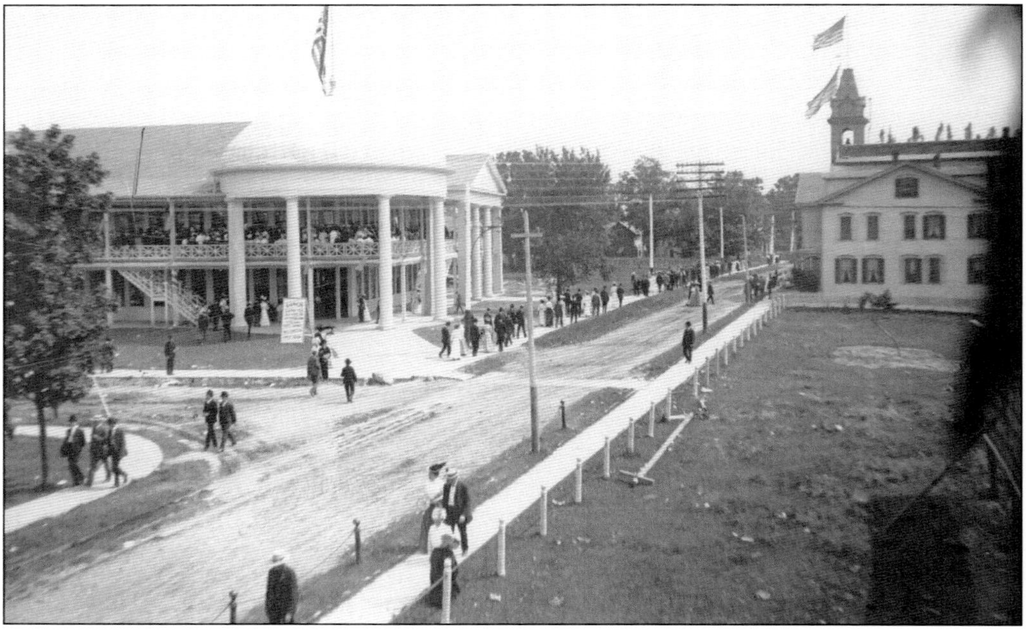

Built in 1905, the Colonial ringed the corner of Delaware and Catawba Avenues in Put-in-Bay, where it served as an iconic landmark for more than 75 years. The Colonial, seen here around 1910, featured a bowling alley, restaurant, bar, dance hall, arcade, movie house, clothing store, and skating rink. The building burned to the ground in May 1988. Today, the Beer Barrel Saloon occupies much of the Colonial's original footprint. (Courtesy of SLARC.)

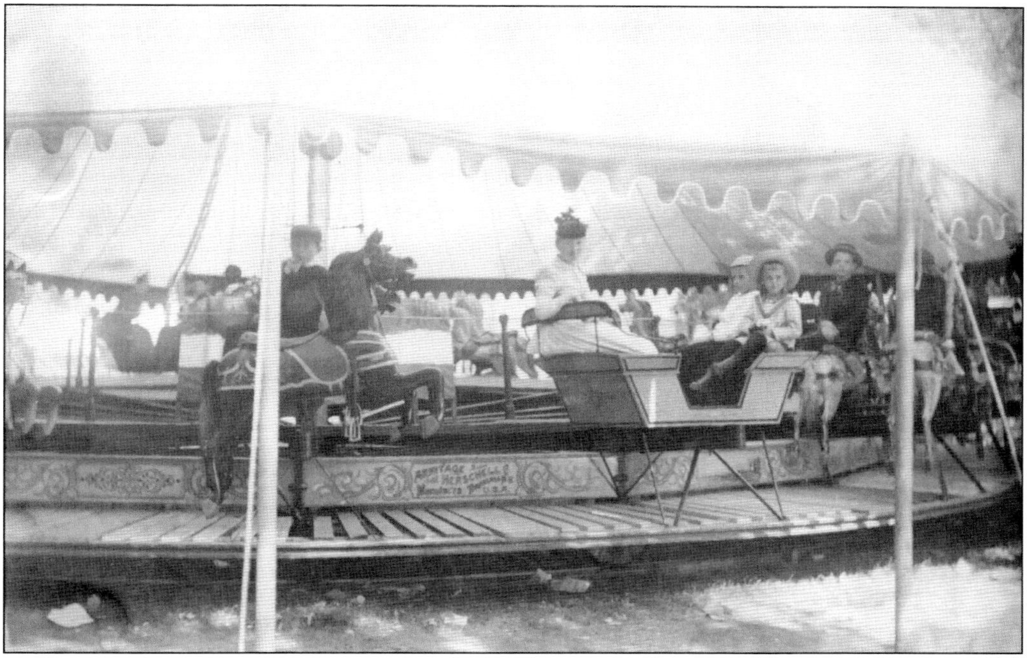

The merry-go-round, or carousel, has been a popular amusement ride since the 19th century. In the summer of 1892, brothers Russell and Henry Cooke Jr. take a ride on the merry-go-round at Put-in-Bay. This carousel was manufactured by the Herschell Company in Tonawanda, New York, one of the country's best-known carousel builders. (Courtesy of HPC.)

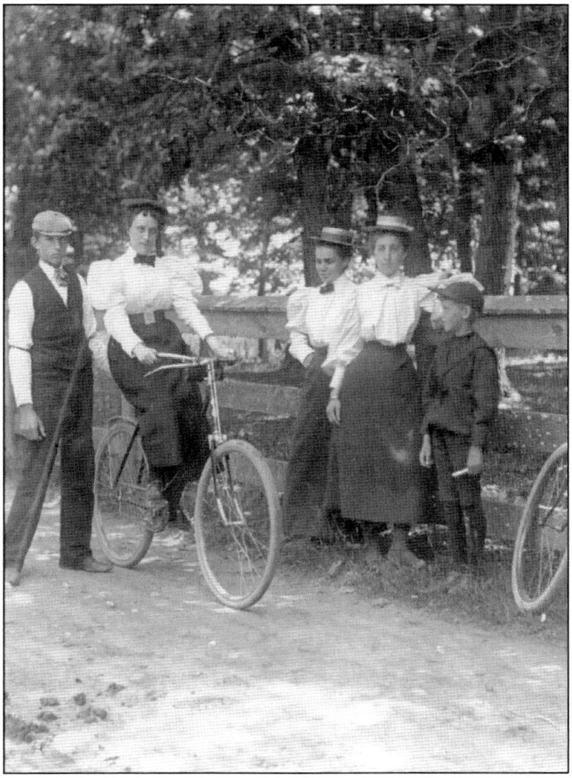

Bicycling was a very popular activity in America in the late 19th and early 20th centuries. It provided transportation, exercise, and social time with friends and family. Caroline (second from right) and Henry Cooke Jr. (far right) watch as Esther Cooke sets out for a ride on South Bass Island in July 1896. (Courtesy of HPC.)

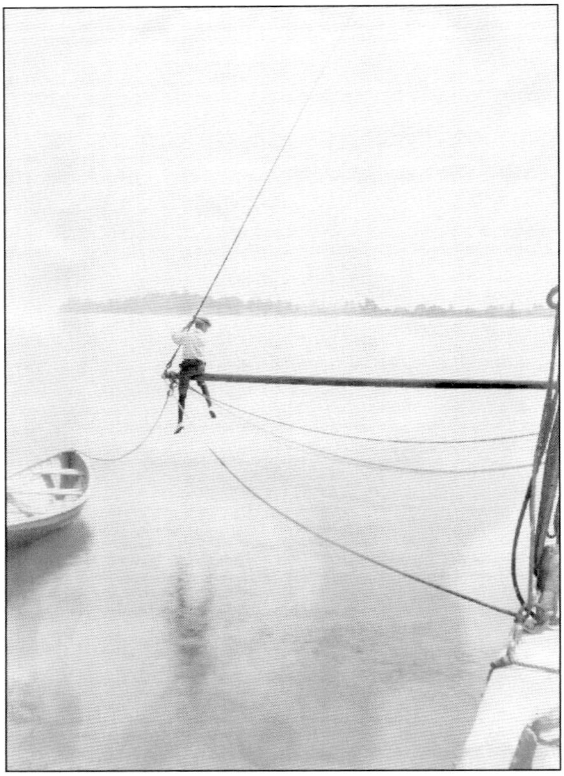

In the summer of 1899, the Reverend Henry Cooke took this photograph of a young man named Gene, a guest of the Cooke family on Gibraltar Island, as he sits on the boom of the ship *Michigan*, which lay at anchor in the bay off Gibraltar. (Courtesy of HPC.)

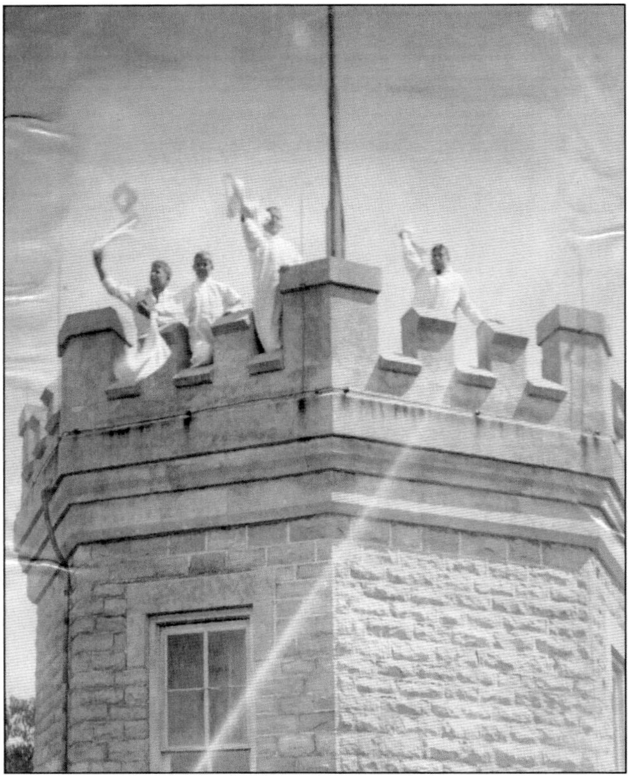

In 1891, several Cooke family members wave to the photographer, the Reverend Henry Cooke, from the tower at the top of Cooke Castle. The roof was reached by a very narrow staircase and trap door. The view of Put-in-Bay Harbor was worth the climb. (Courtesy of HPC.)

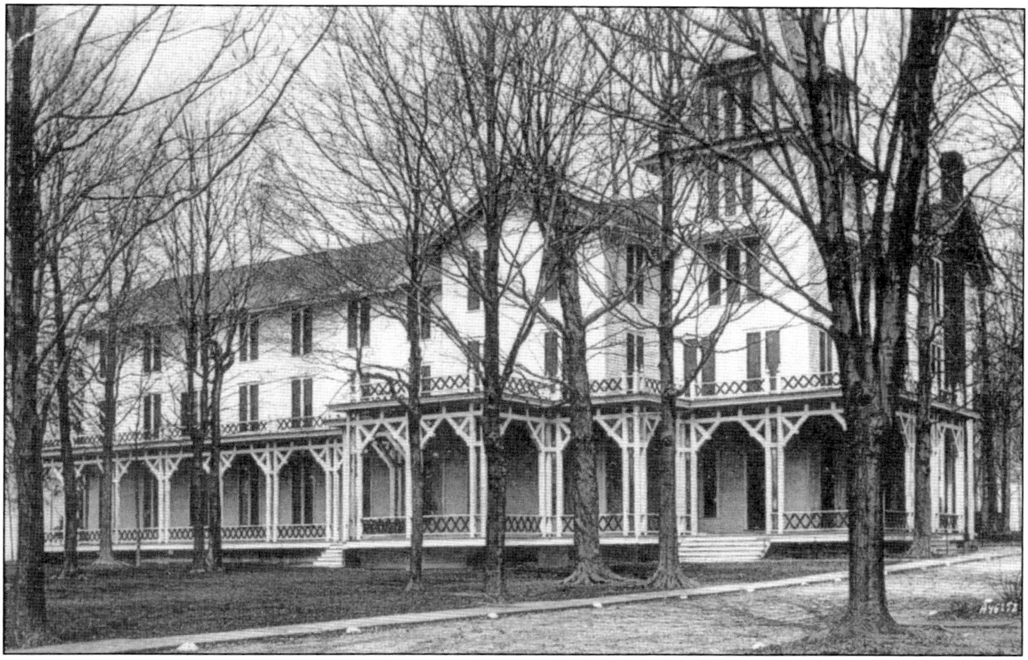

The Middle Bass Island Club, seen here around 1895, attracted visitors such as former US presidents Benjamin Harrison, Grover Cleveland, Rutherford B. Hayes, and William Howard Taft. The club also hosted Ohio politicians and businessmen. They were attracted by the quiet and privacy available on the island. (Courtesy of SAMA.)

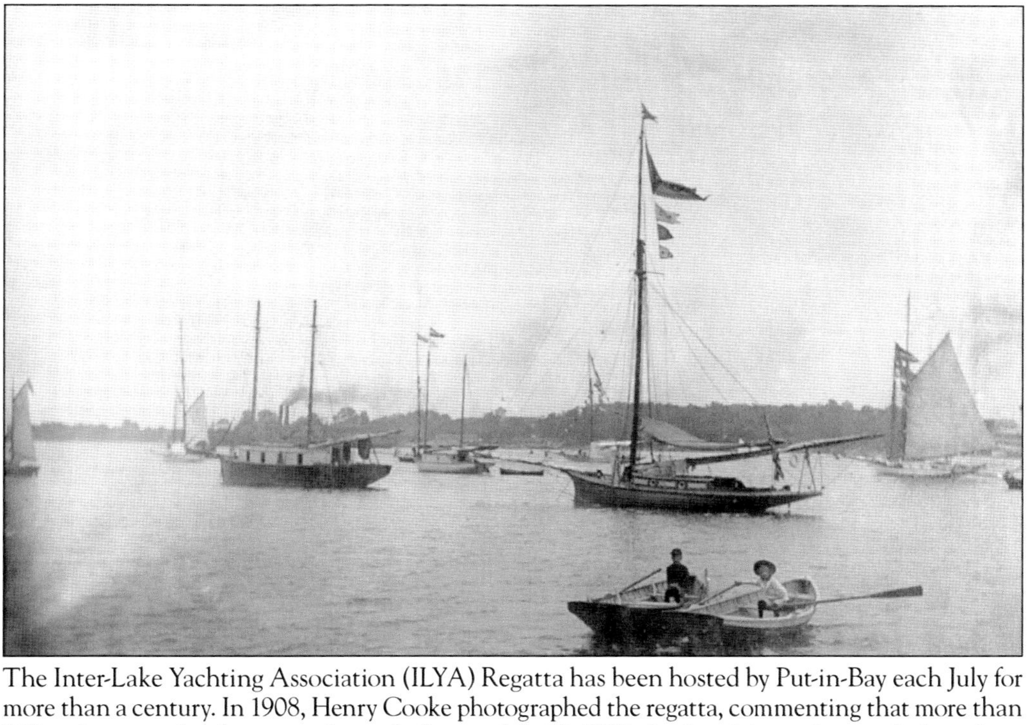

The Inter-Lake Yachting Association (ILYA) Regatta has been hosted by Put-in-Bay each July for more than a century. In 1908, Henry Cooke photographed the regatta, commenting that more than 100 yachts attended the event. A year before, on July 18, 1907, the first ship-to-shore radio broadcast to report race results marked the start of a new era in communications. (Courtesy of HPC.)

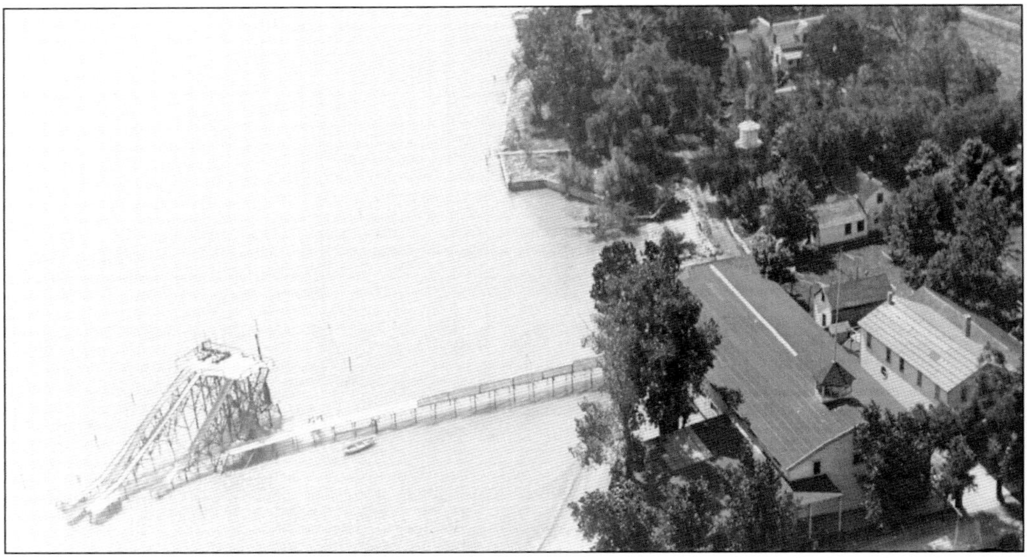

This is a unique view of the toboggan water slide on the South Bass Island bathing beach. Water toboggans were a common sight on Lake Erie through the 1920s. This photograph was taken between 1910 and 1920, probably from Perry's Monument (then under construction) or from an airplane. No comparable view of any other water slide from the period is known to exist. (Courtesy of HPC.)

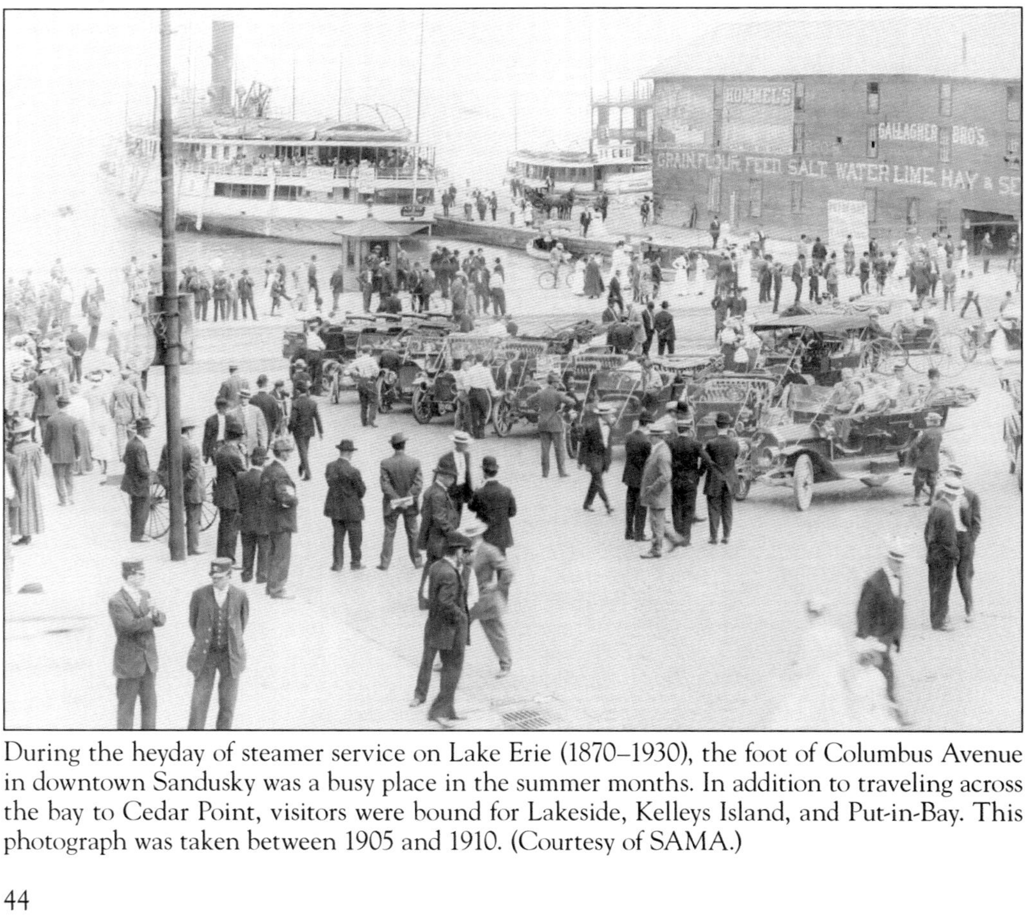

During the heyday of steamer service on Lake Erie (1870–1930), the foot of Columbus Avenue in downtown Sandusky was a busy place in the summer months. In addition to traveling across the bay to Cedar Point, visitors were bound for Lakeside, Kelleys Island, and Put-in-Bay. This photograph was taken between 1905 and 1910. (Courtesy of SAMA.)

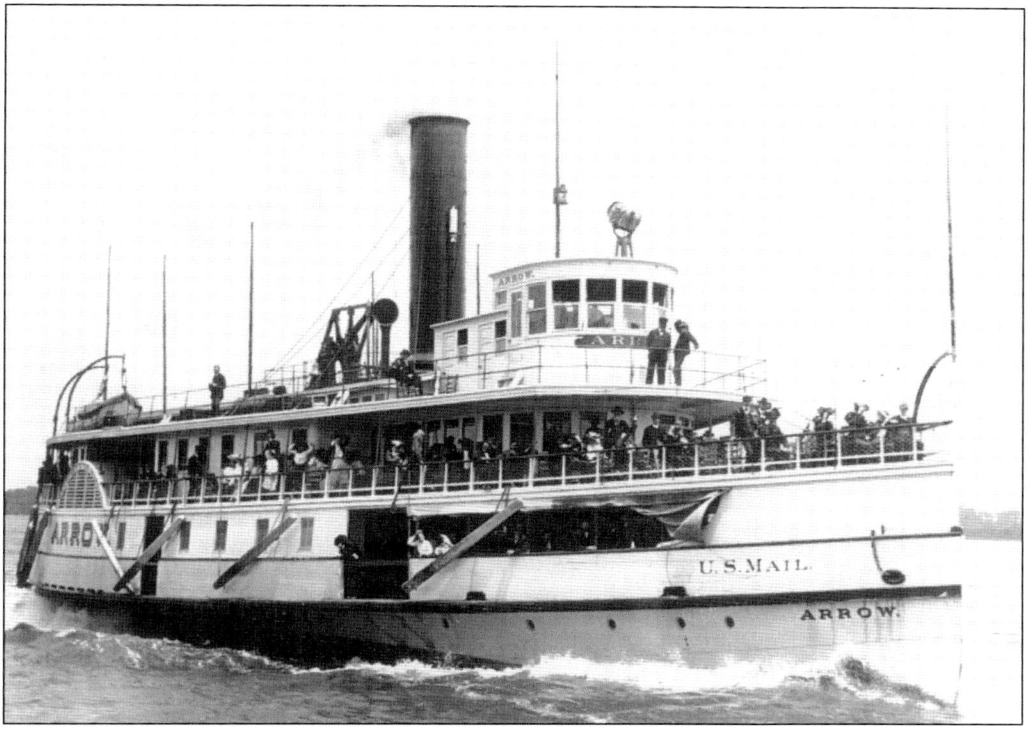

The steamer *Arrow*, seen here at the turn of the century, was built in Wyandotte, Michigan, in 1895 for the Lake Erie trade. The *Arrow*, 165 feet in length, served Sandusky, Cedar Point, Lakeside, Put-in-Bay, and Kelleys for a quarter century, retiring in 1922. Her cabin was furnished in light mahogany, and a grand piano was featured in her main salon. (Courtesy of Frank Landl.)

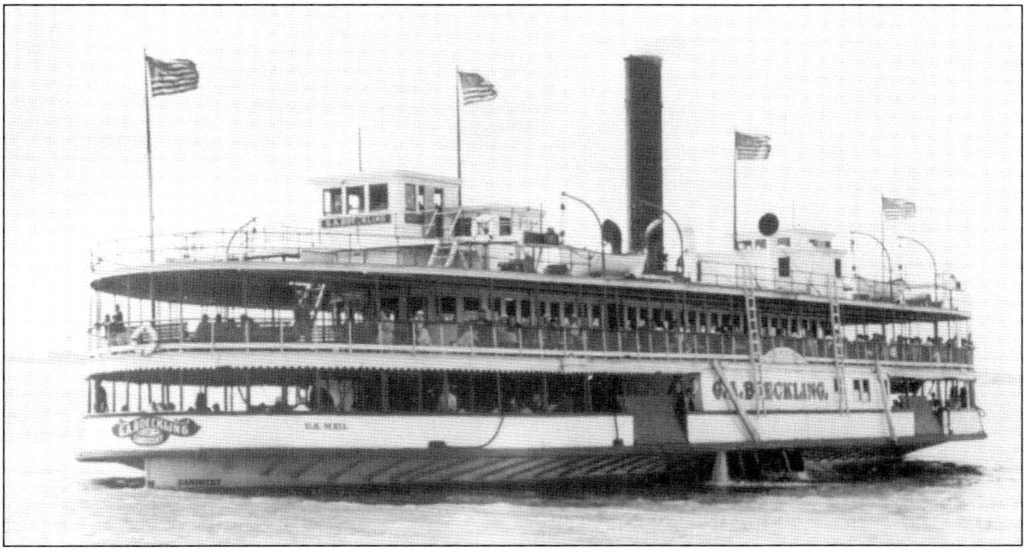

Older Sanduskians still speak fondly about the double-end steamer G.A. *Boeckling*, which carried visitors to and from Cedar Point from 1909 to 1952. After her service to Cedar Point, she was used as a floating warehouse in Wisconsin for many years. In 1983, a group of Sandusky-area residents purchased the boat and returned it to Sandusky. Unfortunately, it was destroyed by fire in 1989 while undergoing restoration in Toledo. The vessel is seen here about 1910. (Courtesy of HPC.)

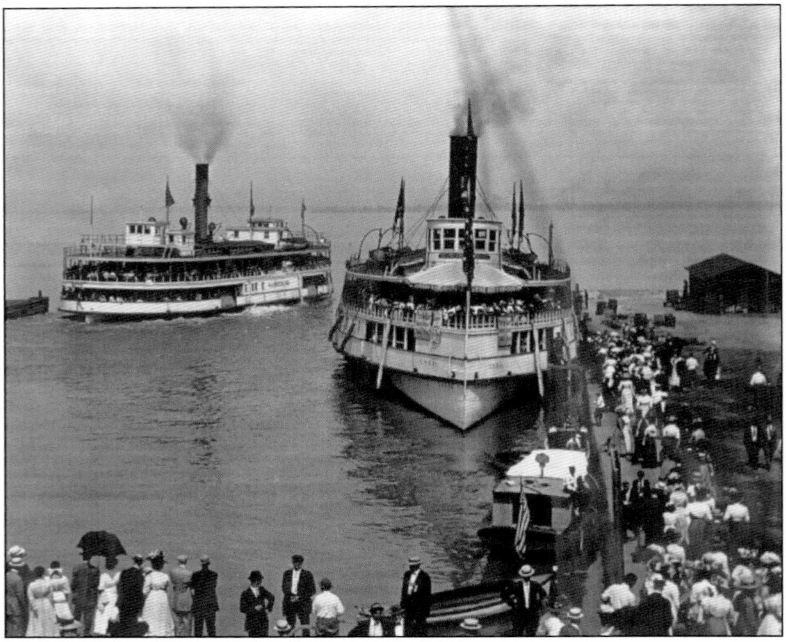

Trains brought many visitors to Sandusky in the summer months, where they could board a waiting steamer for a trip to Cedar Point or Put-in-Bay. Here, around 1910, the G.A. *Boeckling* has just left the dock as passengers quickly board the next boat. (Courtesy of SLARC.)

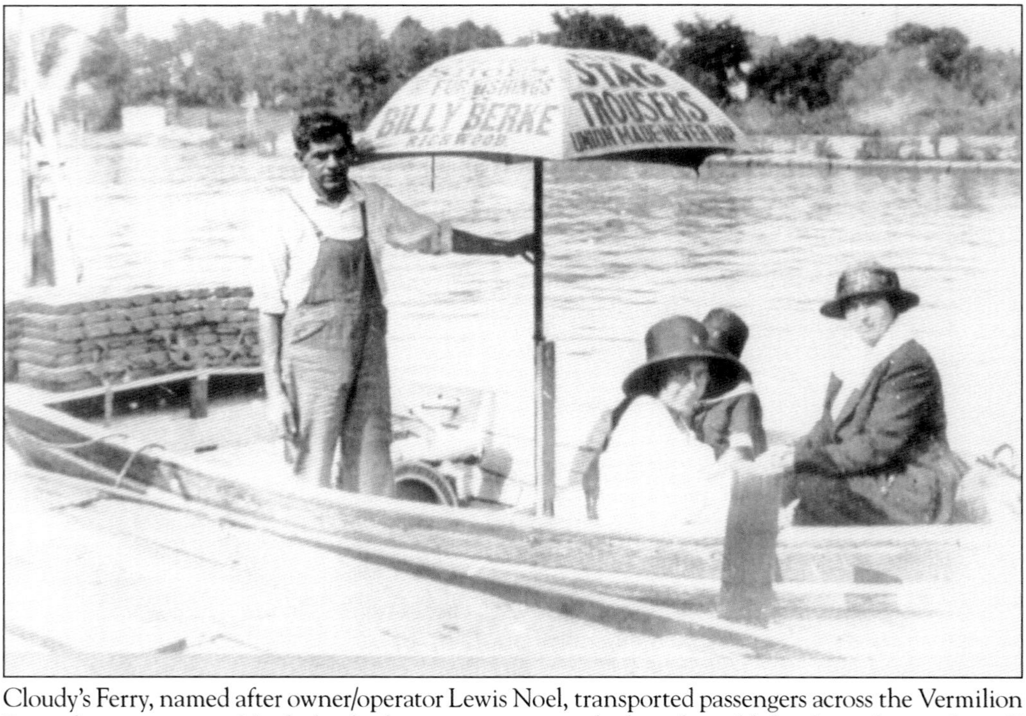

Cloudy's Ferry, named after owner/operator Lewis Noel, transported passengers across the Vermilion River for many years. Noel also had a concession stand where he sold drinks and rented small boats. The ferry is seen here around 1906. (Courtesy of Vermilion Archival Society.)

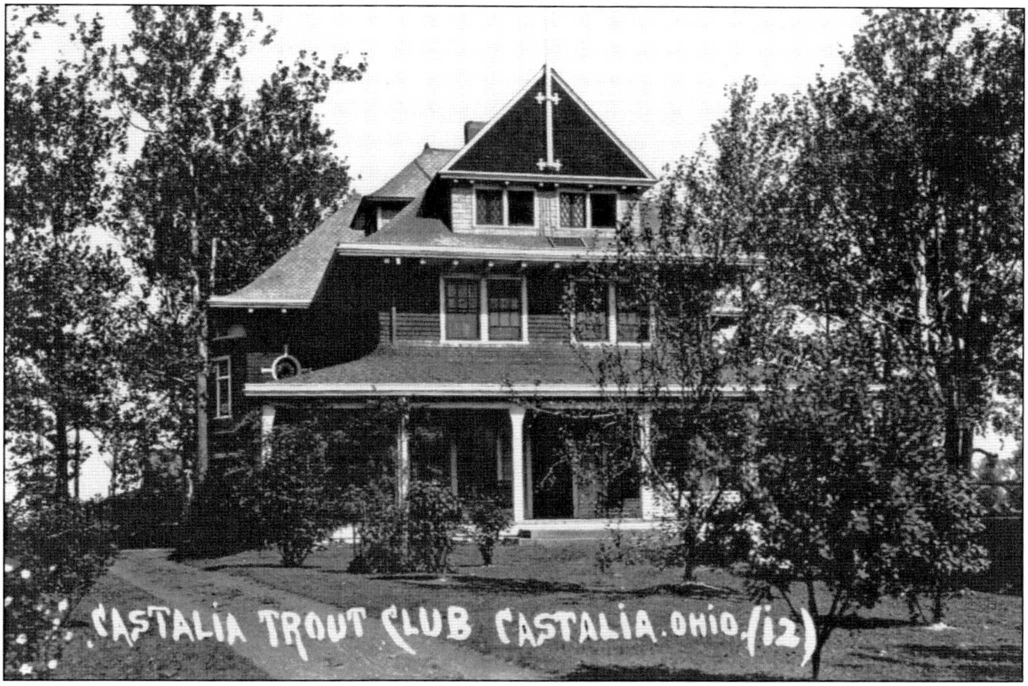

One of many hunting and fishing clubs along the Lake Erie shoreline, the Castalia Trout Club, founded in 1890, provided access to Cold Creek, the only area watercourse with trout habitat. The club is seen here around 1910. (Courtesy of the Castalia Historical Society.)

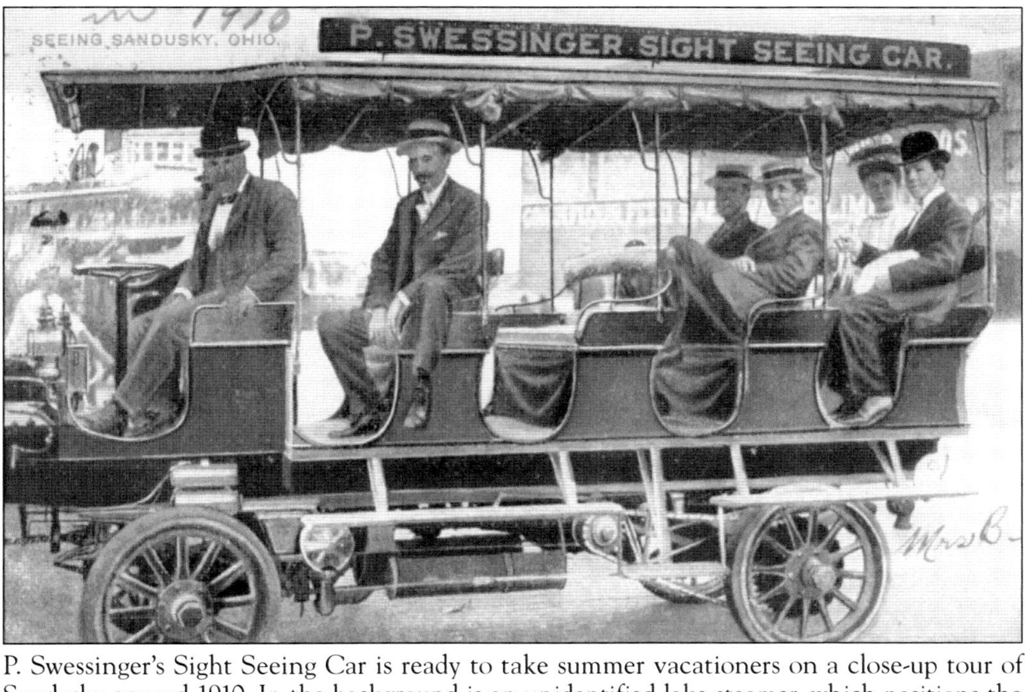

P. Swessinger's Sight Seeing Car is ready to take summer vacationers on a close-up tour of Sandusky around 1910. In the background is an unidentified lake steamer, which positions the car at the foot of Columbus Avenue, the transportation nexus of the Sandusky area. (Courtesy of Frank Landl.)

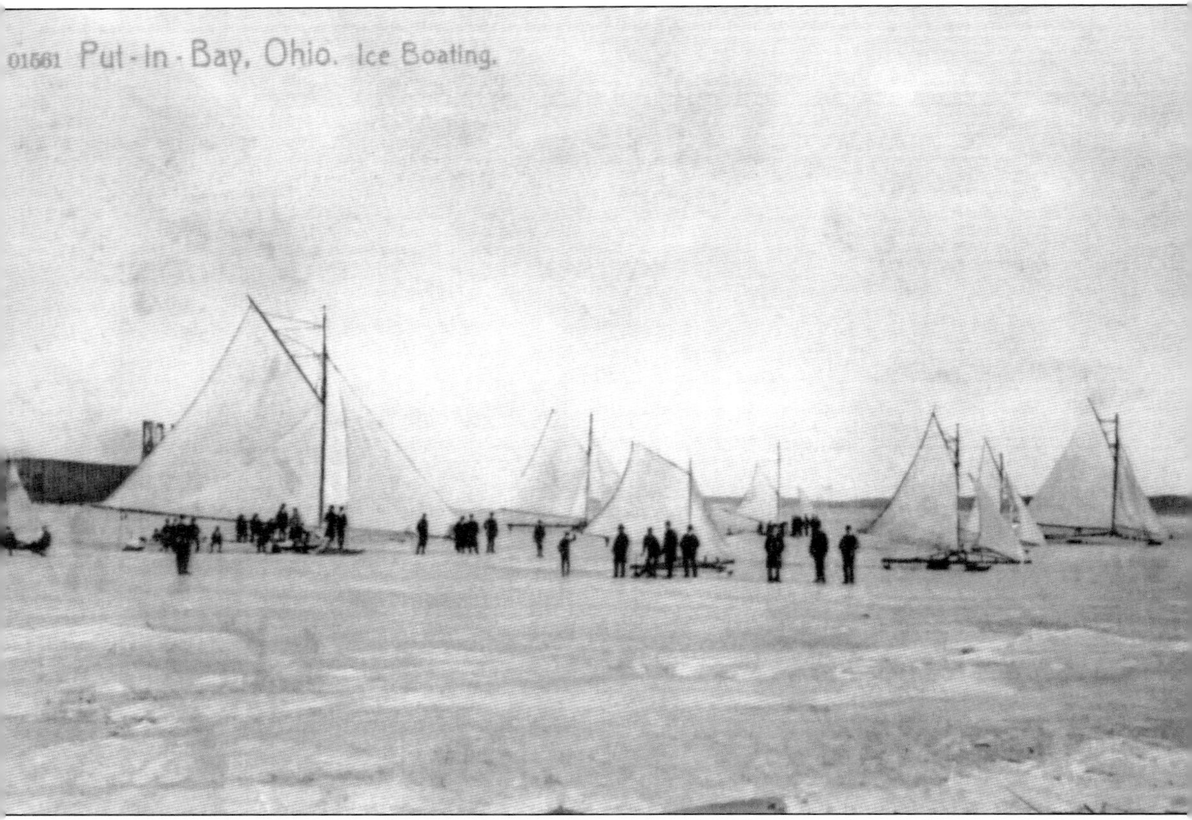

During the winter months, Sandusky Bay was an ideal location for ice boating, attracting enthusiasts from a large area. Gliding across the ice at high speeds was as exhilarating then as it is today. This postcard is from about 1907. (Courtesy of SAMA.)

Three

World War I Era and the 1920s

1911–1931

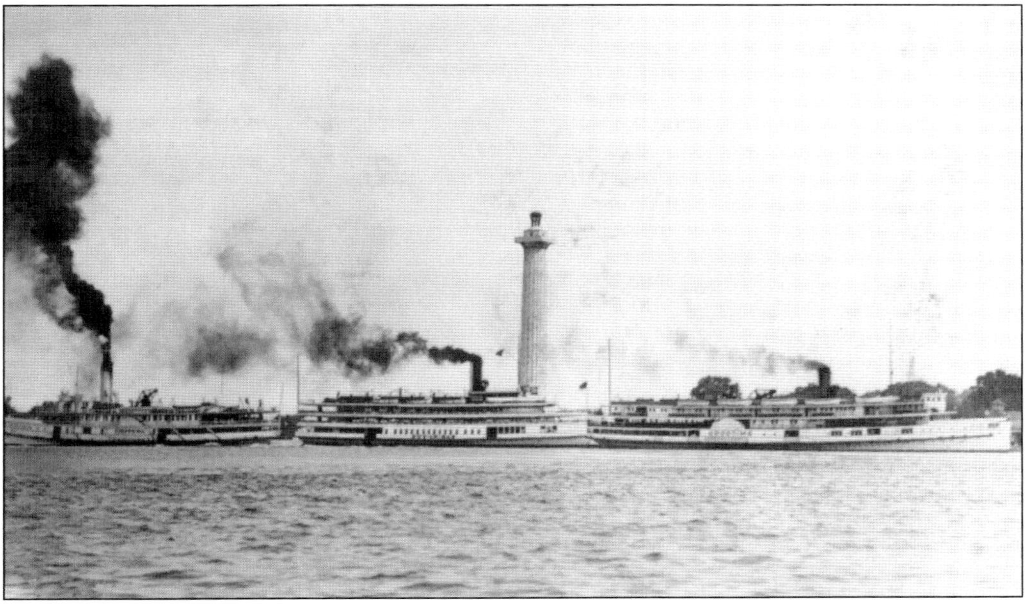

Perry's Victory and International Peace Memorial, built between 1912 and 1915, commemorates Commodore Oliver Hazard Perry's victory over a British fleet in September 1813 and the lasting peace that has existed between England, Canada, and the United States ever since. The Battle of Lake Erie, fought just miles west of Put-in-Bay, was the most significant naval action of the War of 1812. At 352 feet, the monument, seen here around 1918, is the world's tallest Doric column and one of the tallest monuments in the United States. It is visited by 200,000 people each year. (Courtesy of SAMA.)

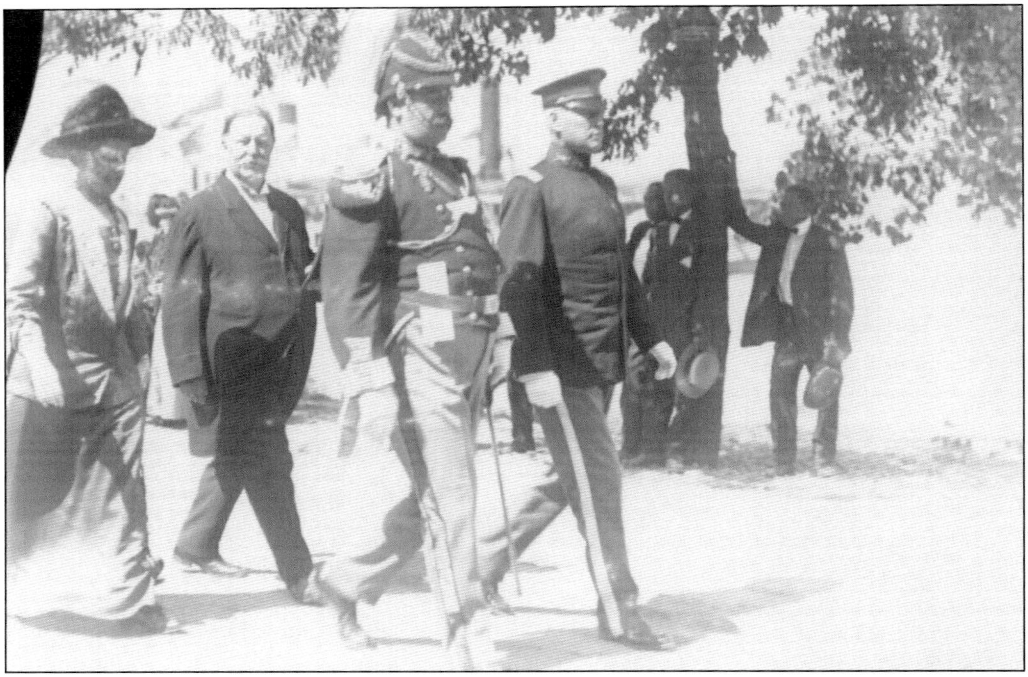

Pres. William Howard Taft (second from left) and his wife, Helen, participate in ceremonies to lay the cornerstone for the monument in August 1912. Initial funding for the monument was coordinated through a multistate commission. While substantially completed by 1915, the federal government did not complete construction of the complex until much later. The official dedication took place on July 31, 1931. (Courtesy of HPC.)

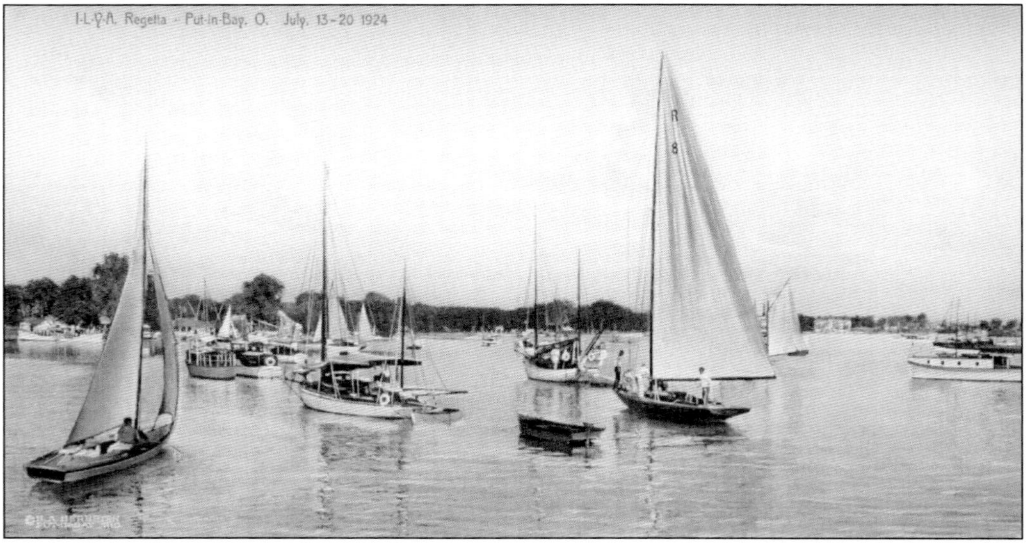

Put-in-Bay Harbor is seen during the 1924 Inter-Lake Yachting Association Regatta. The ILYA Regatta is the oldest event at Put-in-Bay and is still going strong. Primarily a sailing event, it attracts boats from all over the Great Lakes for several days of racing in July or early August. (Courtesy of HPC.)

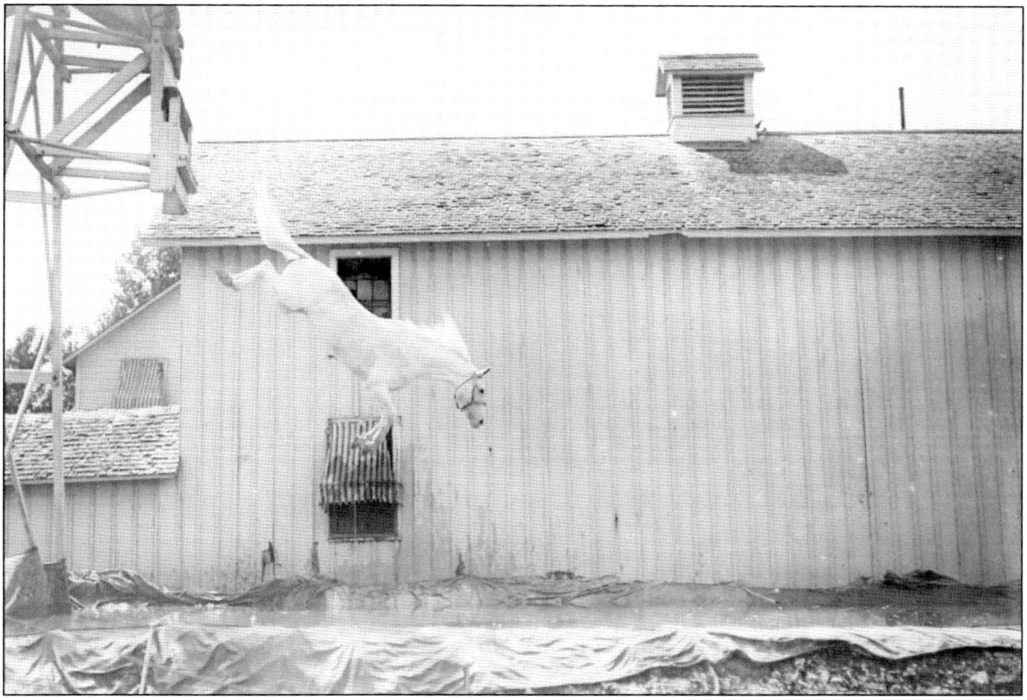

Diving horses were a popular attraction at Put-in-Bay in the 1910s and into the 1920s. The attraction originated in Atlantic City in the 1880s. Horses, such as this one seen around 1920, dove into a pool of water, sometimes from as high as 60 feet. These shows were strongly criticized by animal rights groups, leading to a decline in popularity in the 1930s. (Courtesy of HPC.)

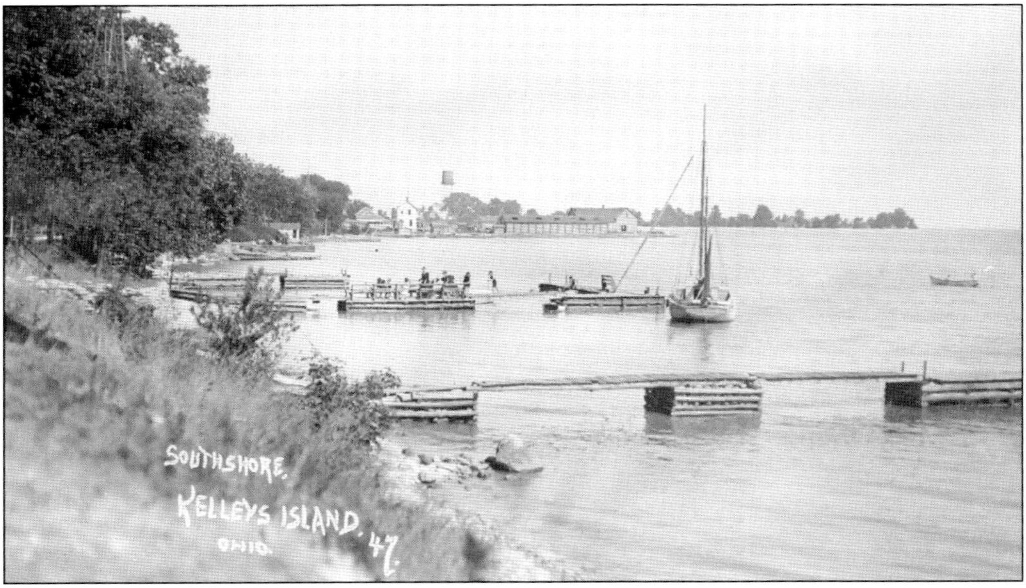

Ernest Niebergall took this photograph around 1918 from the Stone Dock looking east along the south shore of Kelleys Island. Although not as popular as nearby South Bass Island, Kelleys has always attracted its share of summer visitors. Quarrying, commercial fishing, and wine production were major industries on the island for many years, with tourism taking a backseat. (Courtesy of HPC.)

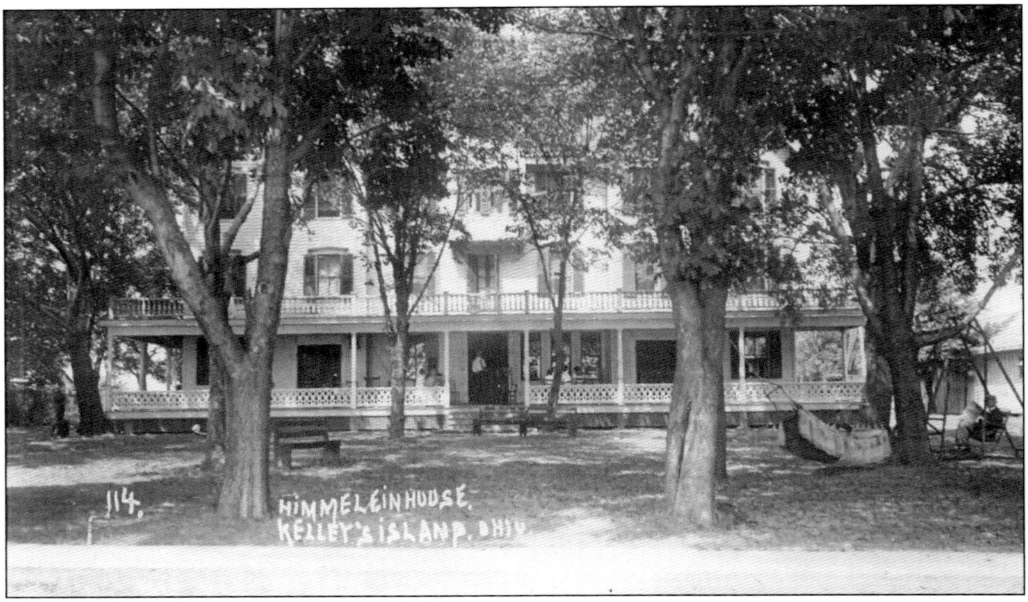

Originally the family home, the Hemmelein Hotel on Kelleys Island welcomed fishermen, grape buyers, and vacationers seeking escape from their busy lives on the mainland. In addition to the hotel, the Hemmeleins later added a dock and beach house. The hotel is seen here around 1916. (Courtesy of HPC.)

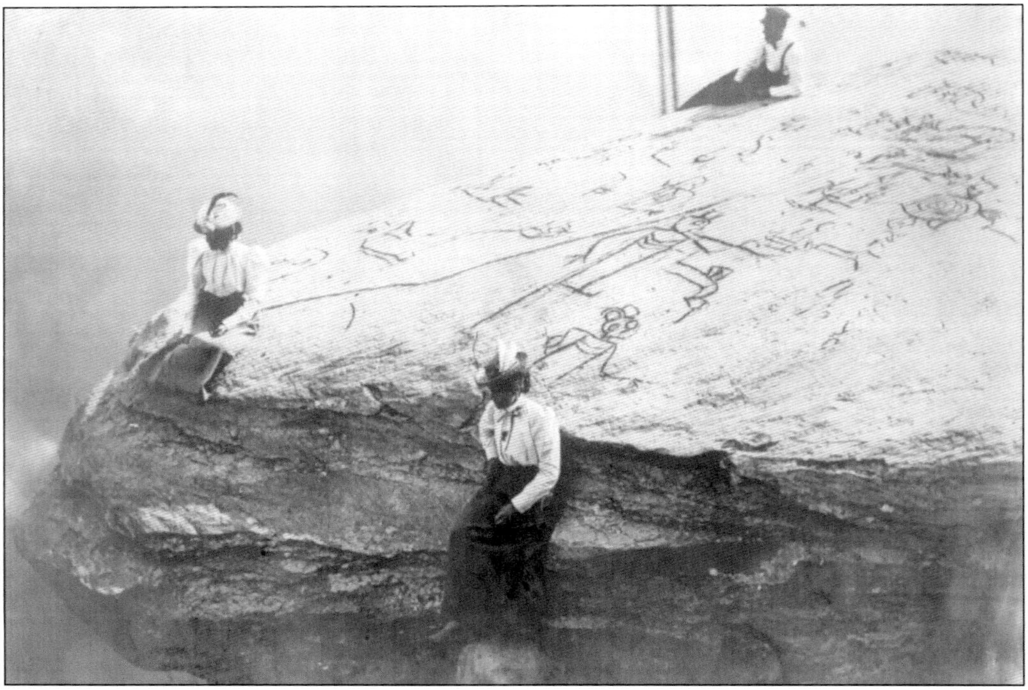

These pictographs on a massive limestone boulder are believed to have been etched by the Erie Indians in the 1600s. Inscription Rock has been a source of interest for Kelleys Island visitors for more than a century. Here, three women perch on the rock to enjoy the lake view around 1915. Today, much diminished by sun, wind, water, and human activity, Inscription Rock is protected by a roofed structure. (Courtesy of HPC.)

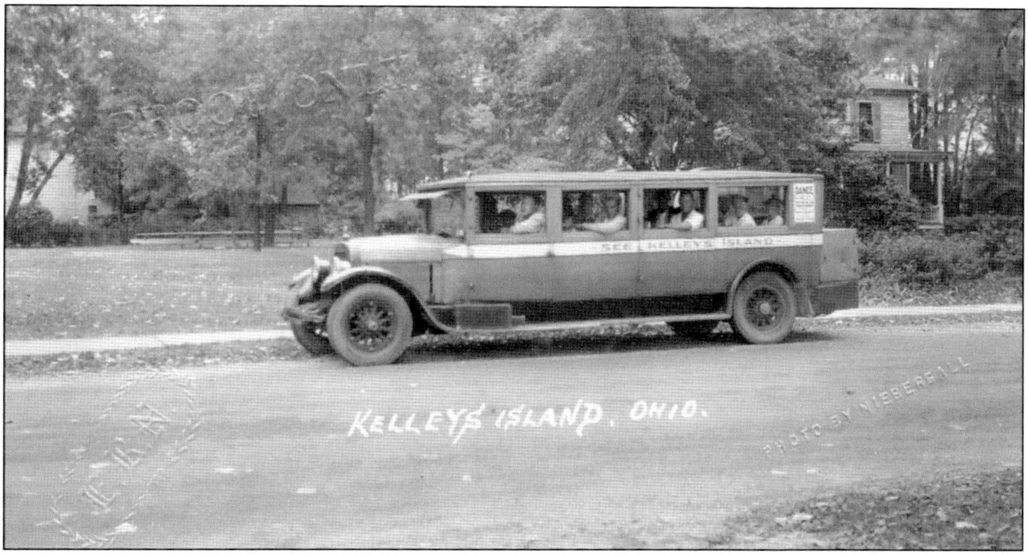

Long before golf cart rentals, a tour bus traveled Kelleys Island offering visitors the opportunity to explore the island from the comfort of the road. The bus also served as a moving billboard, advertising island events; in this case, an island dance around 1918. (Courtesy of HPC.)

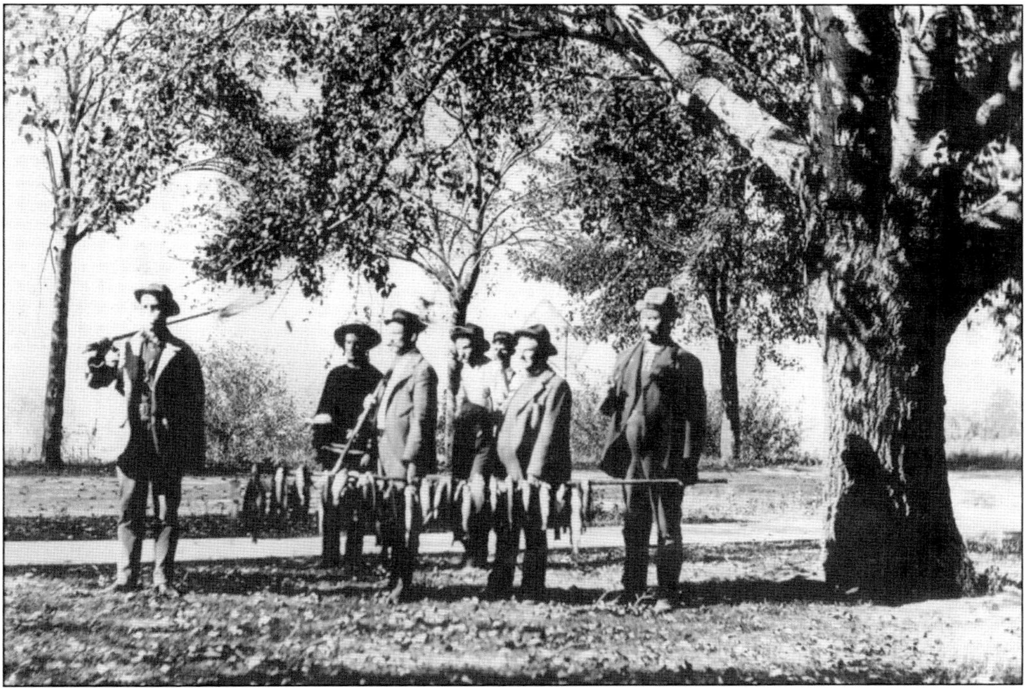

Fishermen display a fine day's catch in the yard of the Schardt Hotel on Kelleys Island around 1914. Sportfishing coexisted with a large commercial fishery in western Lake Erie for more than 75 years. Commercial fishing declined dramatically in the 1950s and 1960s and is now virtually nonexistent on the US side of the lake. The Schardt building still stands and is now a private residence. (Courtesy of HPC.)

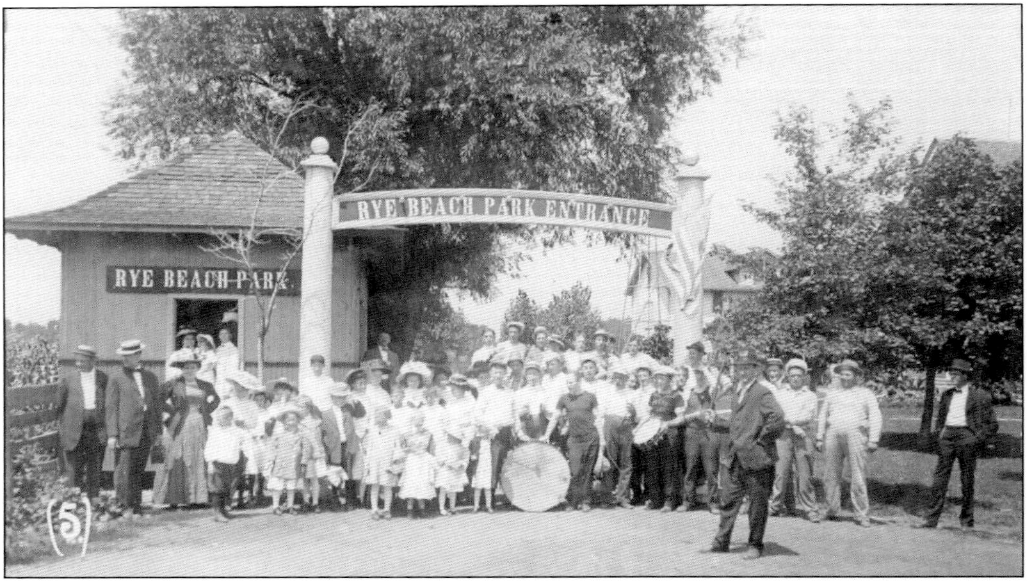

Men, women, and children pose for an Ernst Niebergall photograph at the entrance to Rye Beach Park in the summer of 1913. In the 1910s and 1920s, Rye Beach was a popular spot for picnics and other group activities. (Courtesy of HPC.)

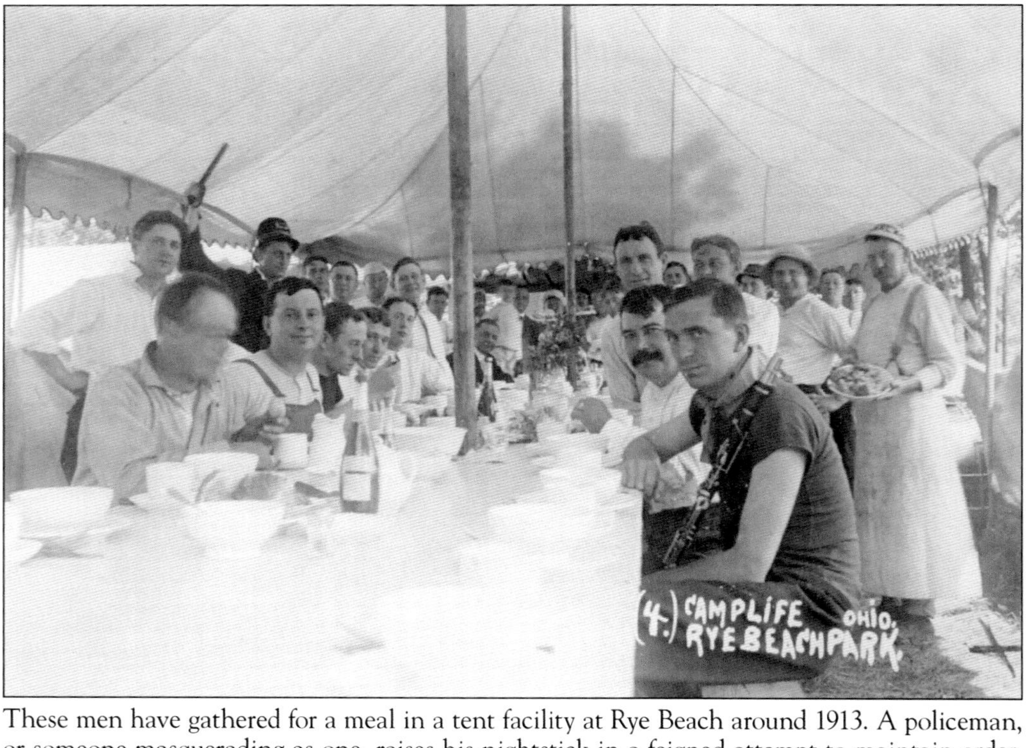

These men have gathered for a meal in a tent facility at Rye Beach around 1913. A policeman, or someone masquerading as one, raises his nightstick in a feigned attempt to maintain order. There are lots of wine bottles on the table. Picnic meals were very elaborate and occupied much of the day. (Courtesy of HPC.)

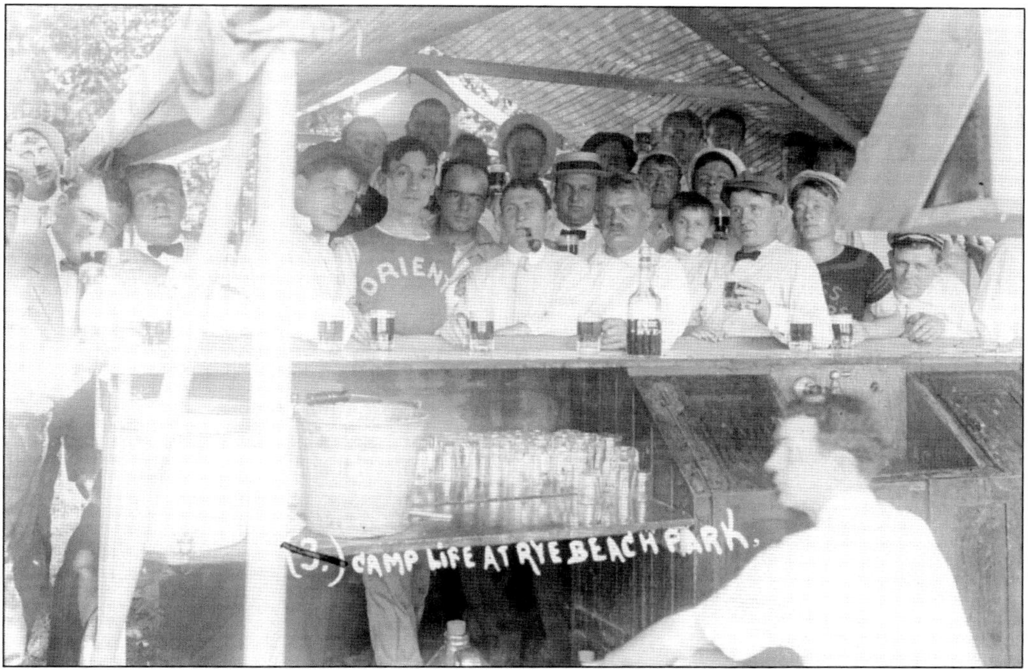

William Bruns, owner of Rye Beach Park, did not flinch from serving alcohol to his guests. Drinking was a popular activity at the park, at least until the advent of Prohibition in 1919. These men enjoy some of the "pure rye" that was served at the park around 1913. (Courtesy of HPC.)

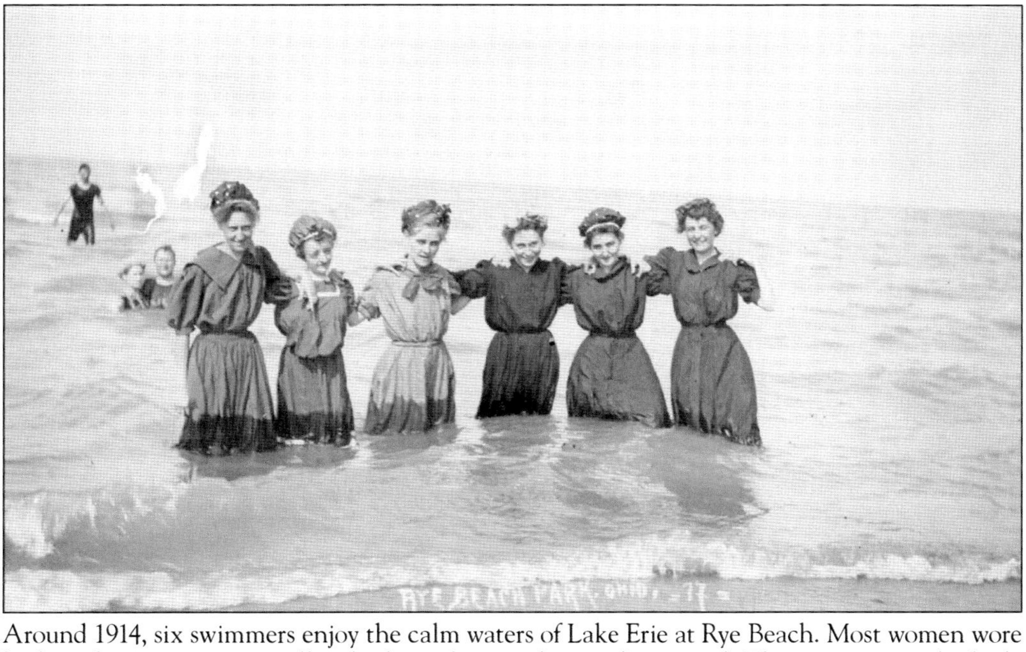

Around 1914, six swimmers enjoy the calm waters of Lake Erie at Rye Beach. Most women wore bathing hats or caps, as well as bathing dresses, during this period. There appears to be little variation among the dresses. A couple in the water behind the group carefully observes the scene. (Courtesy of HPC.)

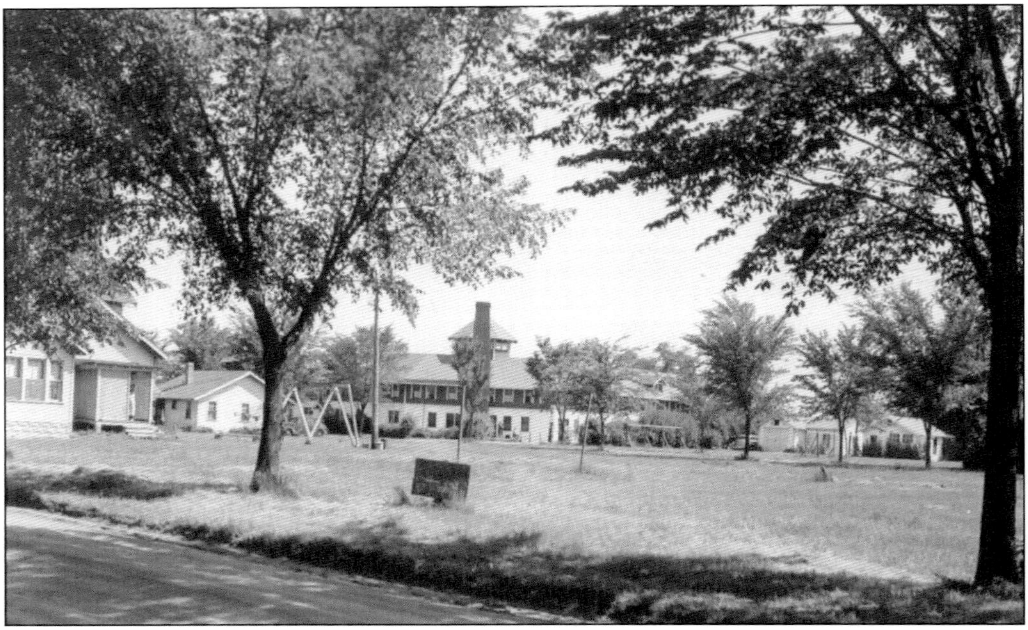

Beulah Beach, founded in 1920 by the Christian and Missionary Alliance, was one of several religious communities that grew up on the lakeshore between Vermilion and Sandusky. In 1926, the Beulah Beach Bible Institute was opened to train ministers and missionaries. Cottages were built on the property, and a large barn was converted to a tabernacle. This photograph was taken around 1920. (Courtesy of HPC.)

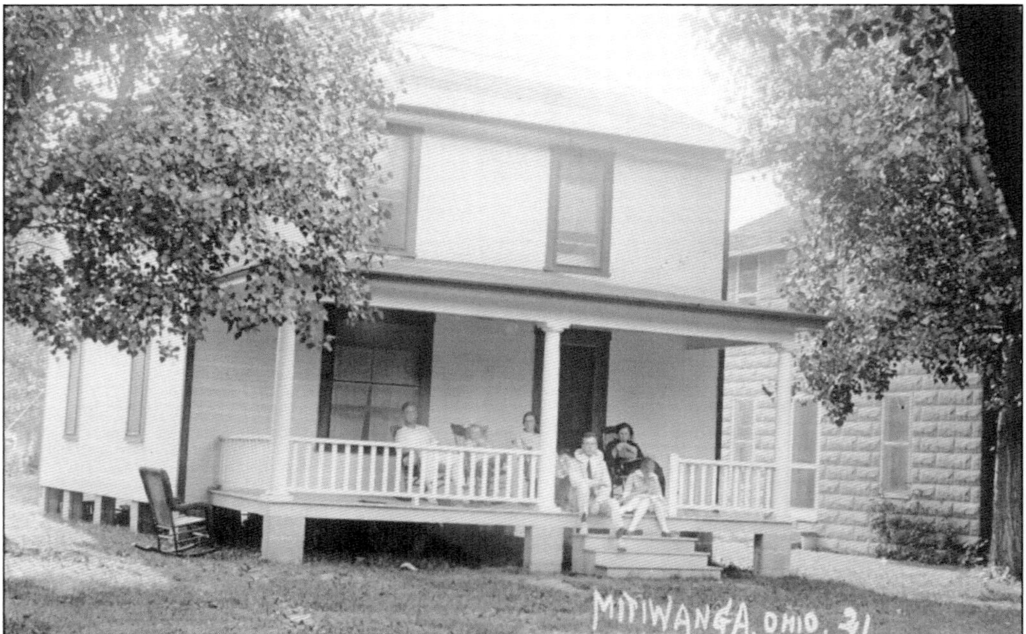

Vacationers of all ages enjoy the porch at this cottage in Mitiwanga Park in 1912. According to local legend, the name Mitiwanga derives from a Native American word meaning "peaceful shore." Its growth as a cottage community was encouraged by the development of the interurban railway from Cleveland to Sandusky in 1900. (Courtesy of HPC.)

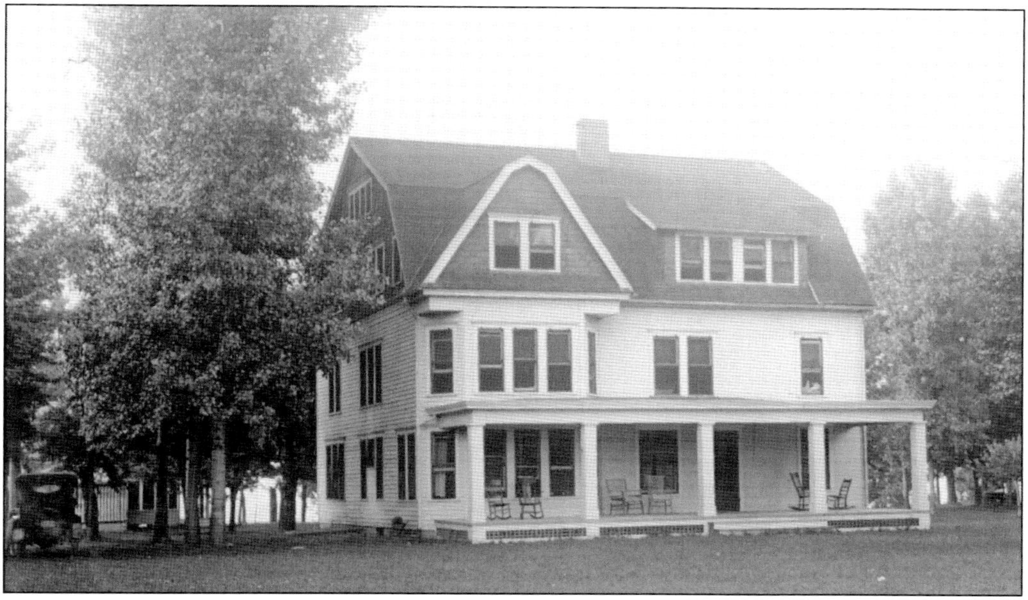

The Ruggles Grove property was originally owned by pioneer Almon Ruggles, who received the land as compensation for surveying the "Firelands" for the government of Connecticut in the early 1800s. Almon's son Richard opened the grove for public picnics in 1868. Richard's son Charles changed the name to Ruggles Beach and opened it as a summer resort in 1891, with cottages, a hotel, and a swimming beach. Ruggles Hotel is seen here sometime between 1910 and 1915. (Courtesy of HPC.)

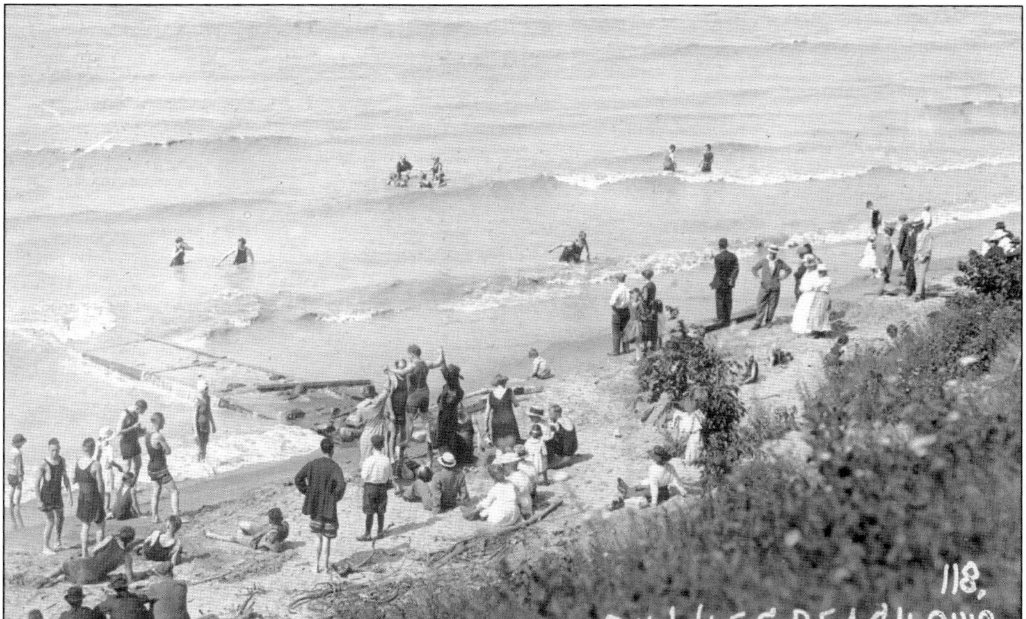

Just left of center in this c. 1912 photograph, two women help a young man keep his balance as he does an acrobatic stand at Ruggles Beach. The photographer, Ernst Niebergall, positioned his camera on the bluff above the beach. Interestingly, the man doing the acrobatics was probably not aware he was being photographed. (Courtesy of HPC.)

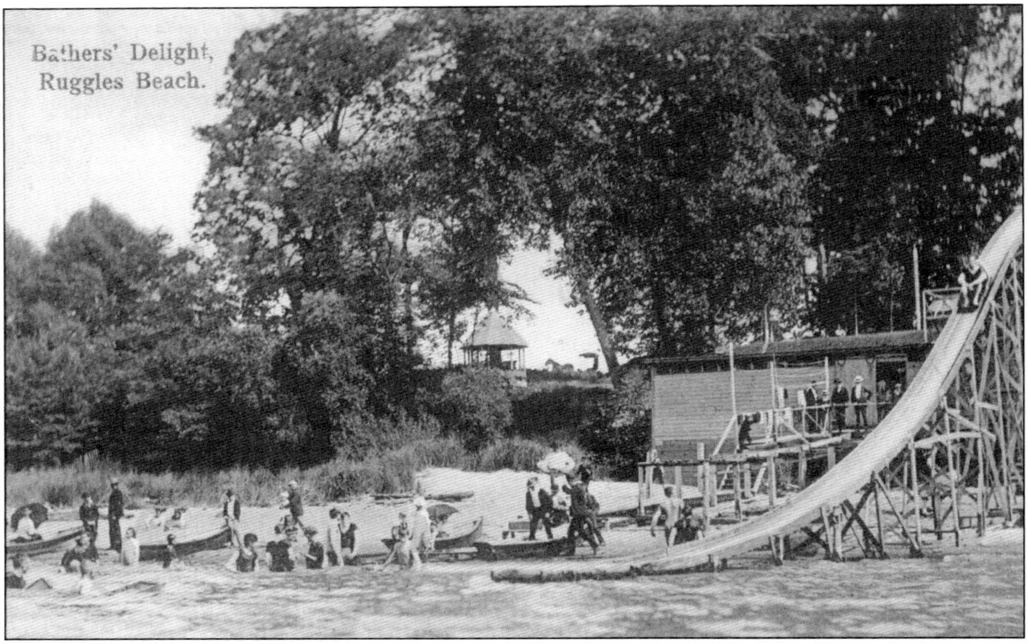

The Ruggles Beach version of the popular water toboggan ride is seen here around 1920. Most Lake Erie beach resorts or parks had some kind of a water toboggan. In this photograph, a rider on a sled has started to make the slide into Lake Erie. (Courtesy of Vermilion Archival Society.)

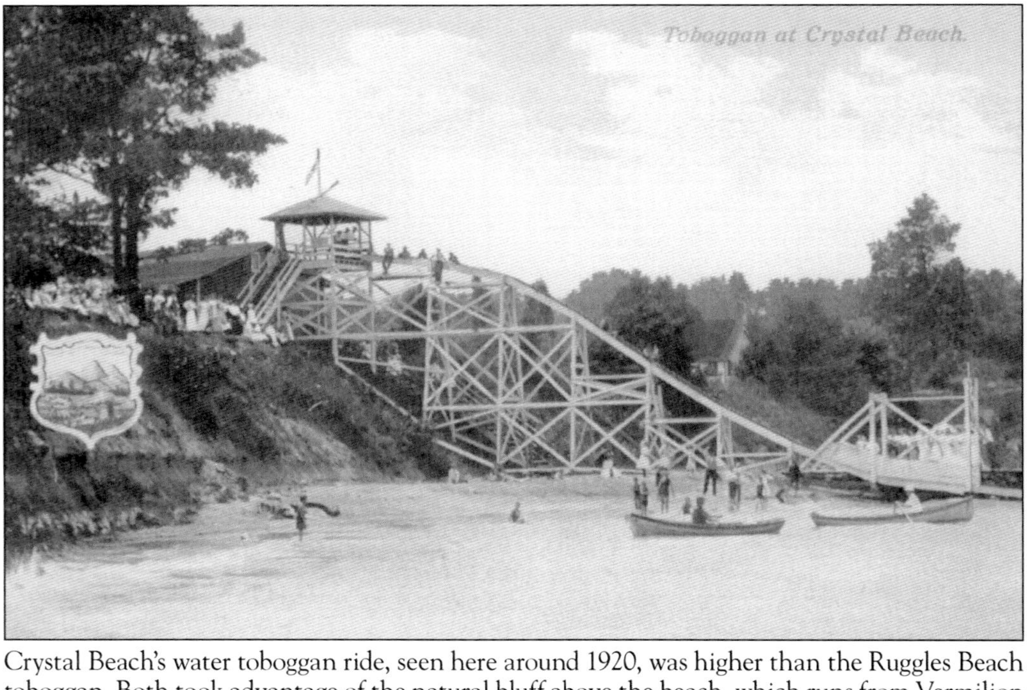

Crystal Beach's water toboggan ride, seen here around 1920, was higher than the Ruggles Beach toboggan. Both took advantage of the natural bluff above the beach, which runs from Vermilion to Huron. Some water toboggans used steam power to help lift individual toboggans back up to the top of the structure. (Courtesy of Vermilion Archival Society.)

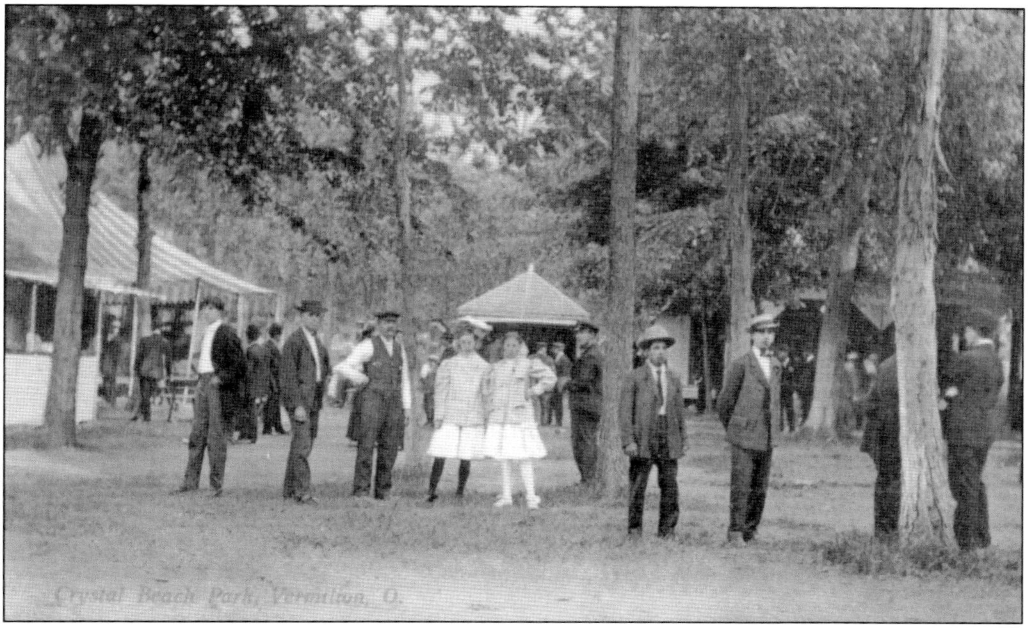

The Crystal Beach story starts with a pioneer farmer, George Shadduck, who in 1874 began converting farmland just east of the Vermilion River into a picnic grove, beer garden, dance hall, and swimming beach. He later sold the property to George H. Blanchat, who added dramatically to the attraction's offering and opened Crystal Beach Park in 1907 as an amusement park. This photograph was taken around 1925. (Courtesy of Vermilion Archival Society.)

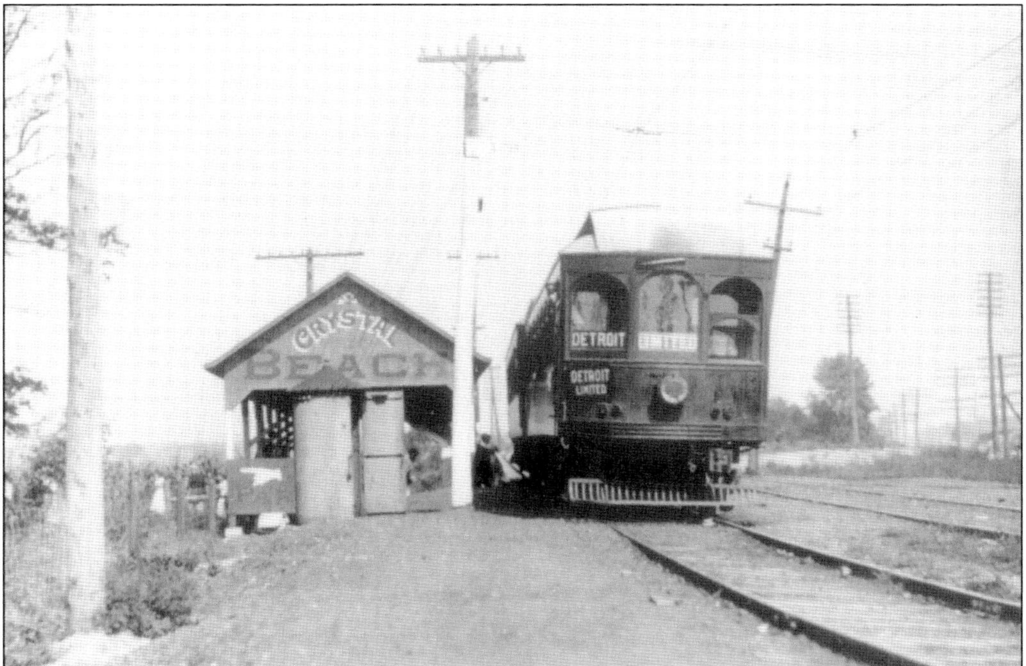

One of the keys to Crystal Beach's success was the Lake Shore Electric Railway, which ran from Cleveland to Detroit, with a stop at the park. The growing population in Cleveland had easy access to the park. This photograph was taken around 1920. (Courtesy of Vermilion Archival Society.)

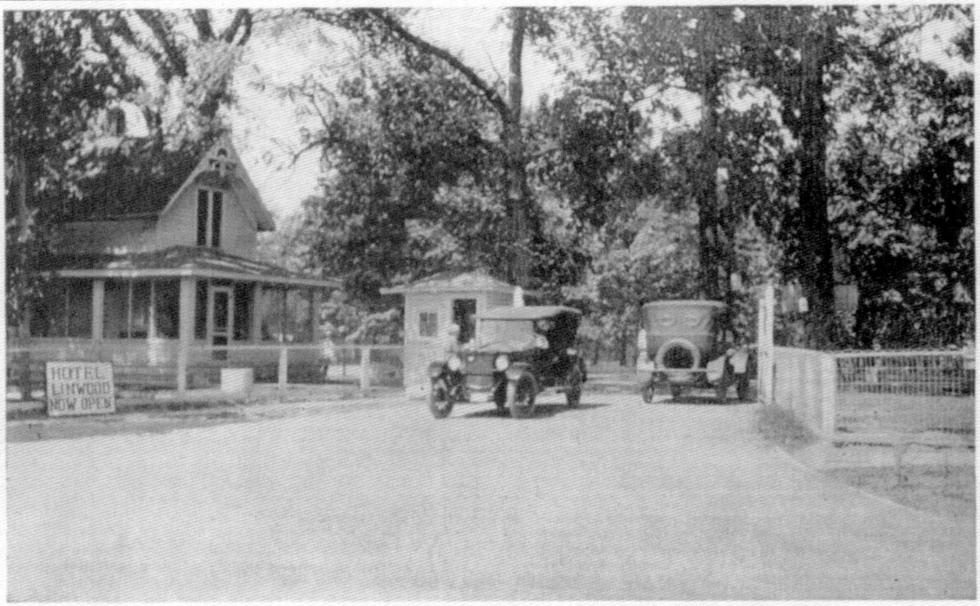

ENTRANCE TO LINWOOD PARK AND SUPERINTENDENT'S RESIDENCE, VERMILION, OHIO.

The automobile entrance to Linwood Park is seen here around 1920. The superintendent's house was also located at the main access point to the park. Linwood Park and Crystal Beach Park were next door neighbors, located just east of the Vermilion River. Linwood Park is today operated as a gated community during the summer months. (Courtesy of Vermilion Archival Society.)

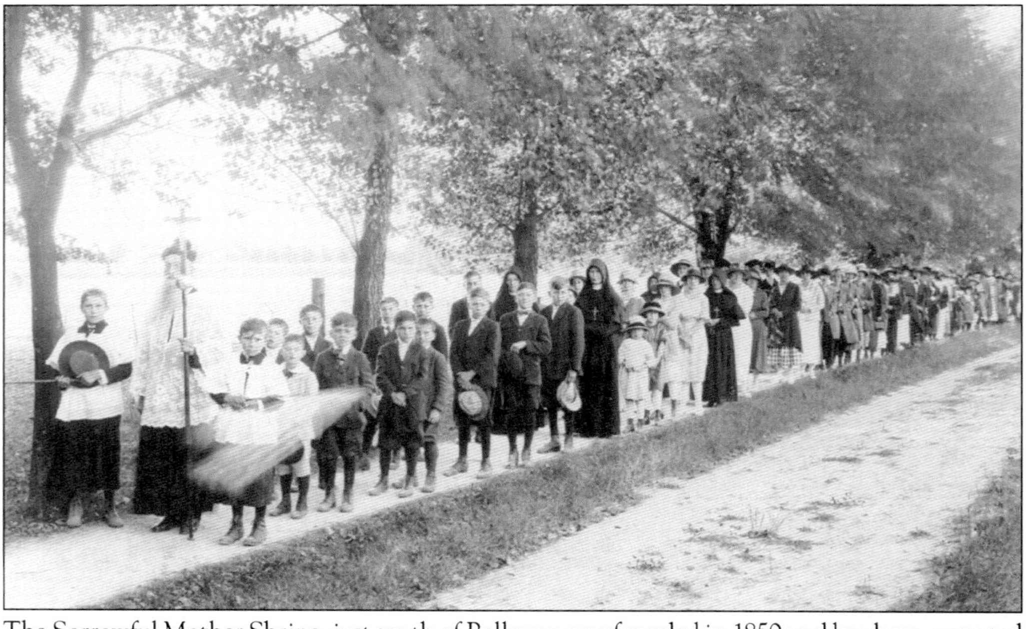

The Sorrowful Mother Shrine, just south of Bellevue, was founded in 1850 and has been operated throughout its history by the Catholic order of the Missionaries of the Precious Blood. The shrine is the oldest place of pilgrimage dedicated to the Blessed Mother in the Midwest and east of the Mississippi River. For more than a century, thousands of Catholics have made pilgrimages to the shrine, which is located in a beautiful 140-acre forest. Here, a priest leads a group of pilgrims to the shrine around 1910. (Courtesy of Sorrowful Mother Shrine.)

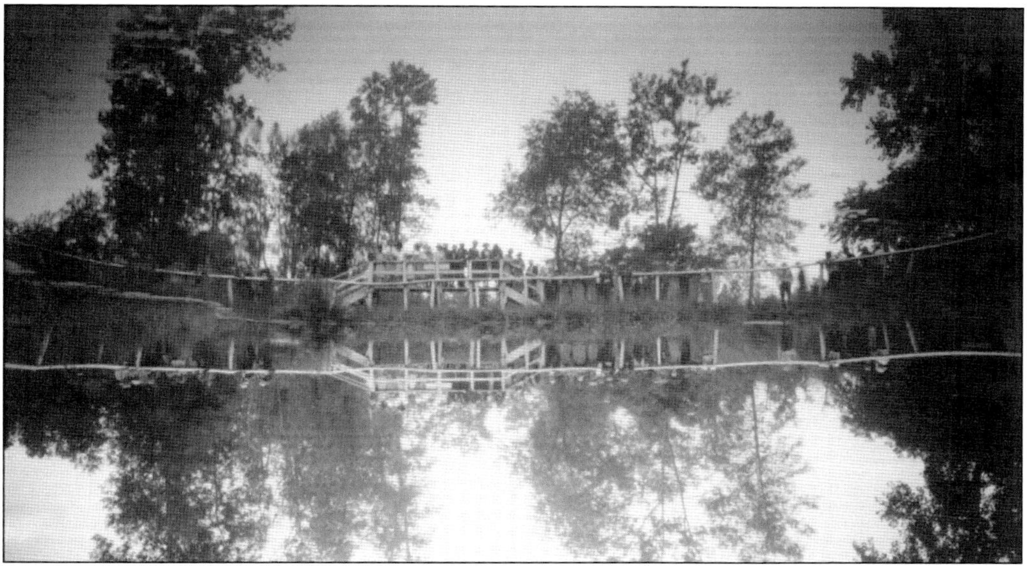

The Blue Hole in Castalia became a popular tourist spot in the 1920s and operated until 1990, attracting 165,000 visitors annually at its peak in the 1970s. It is fed by an underground stream carrying seven million gallons of water daily into Sandusky Bay and Lake Erie. It caught the public's fancy because of its size (about 75 feet in diameter), clarity, vibrant bluish color, and supposed "unknown depth." In actuality, the depth is about 48 feet. The Blue Hole, seen here around 1925, was known to Native Americans; its first recorded mention came in 1761. (Courtesy of HPC.)

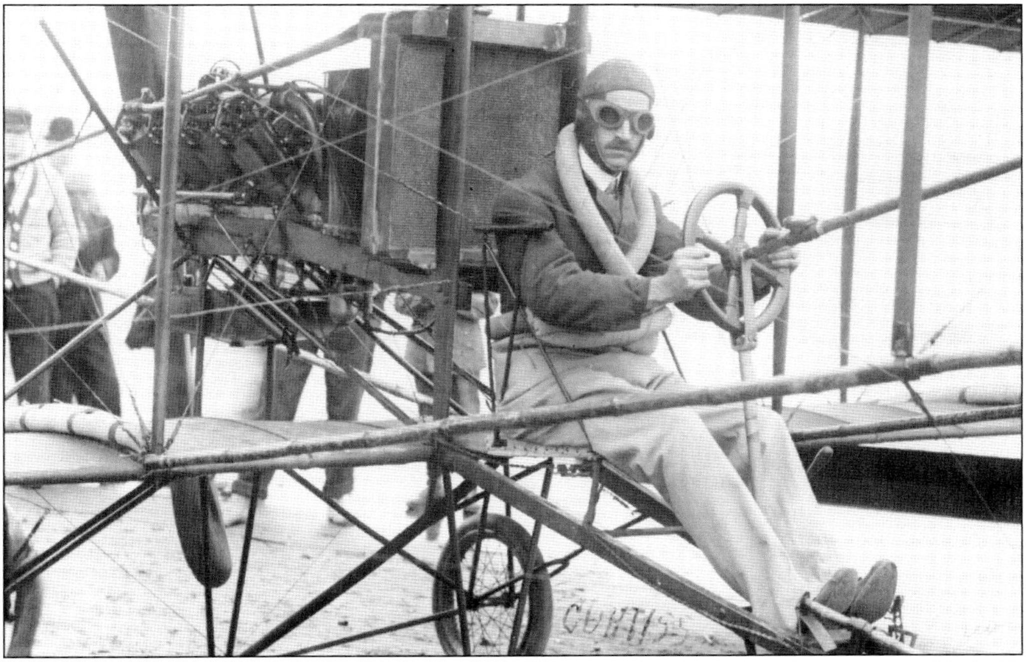

Pioneer aviator Glen Curtiss landed his biplane on the Cedar Point Beach on August 31, 1910, setting world records for altitude (1,000 feet) and distance over water (65 miles). He began his flight at Euclid Beach Park, just east of Cleveland. A crowd of 10,000 spectators, including George Boeckling, was on hand to welcome Curtiss. He wore a bicycle tire as a life vest. (Courtesy of CPA.)

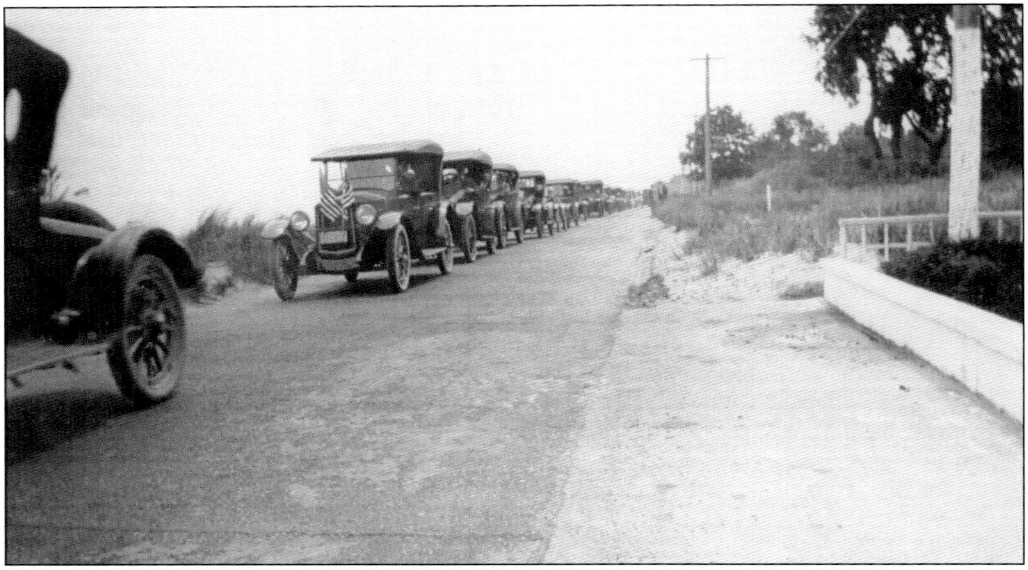

George Boeckling recognized that for Cedar Point to continue to prosper it would need direct access by automobile. In 1914, he opened the Cedar Point Chaussee (*chaussee* is French for a paved road). Seen here some time between 1914 and 1918, the road ran the length of the peninsula, from US Route 6 to the front door of the park, a distance of seven miles. Although an expensive undertaking, it was a necessary one. Boeckling was a visionary, and he understood the impact of the automobile on the park. (Courtesy of HPC.)

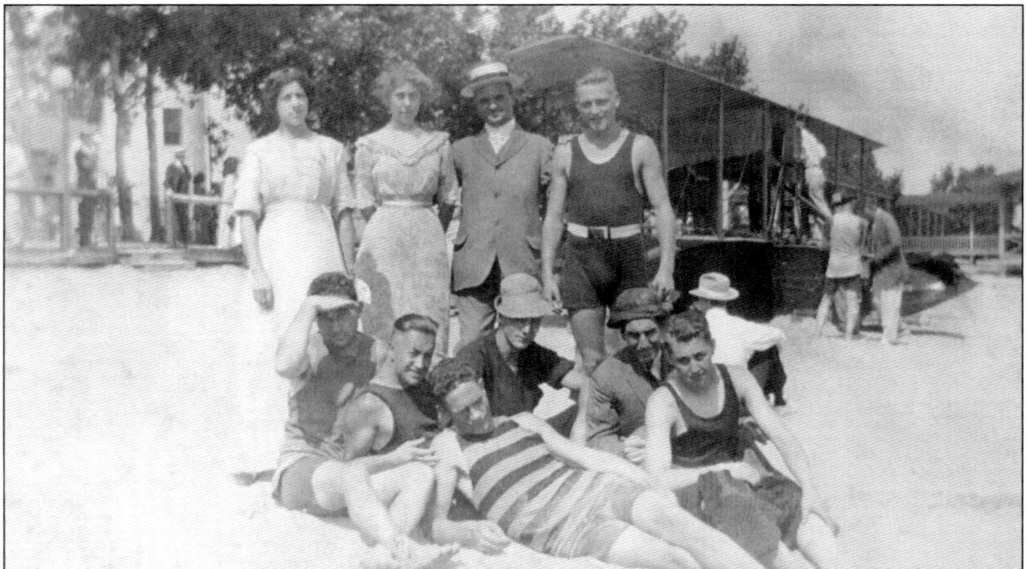

Young people pose for a photograph on the Cedar Point Beach around 1918. Some are wearing swimwear, and others sport formal attire. A "flying boat" is in the background. These airplanes were often used as a prop for souvenir pictures on the Cedar Point Beach. The Hotel Breakers is in the background. (Courtesy of CPA.)

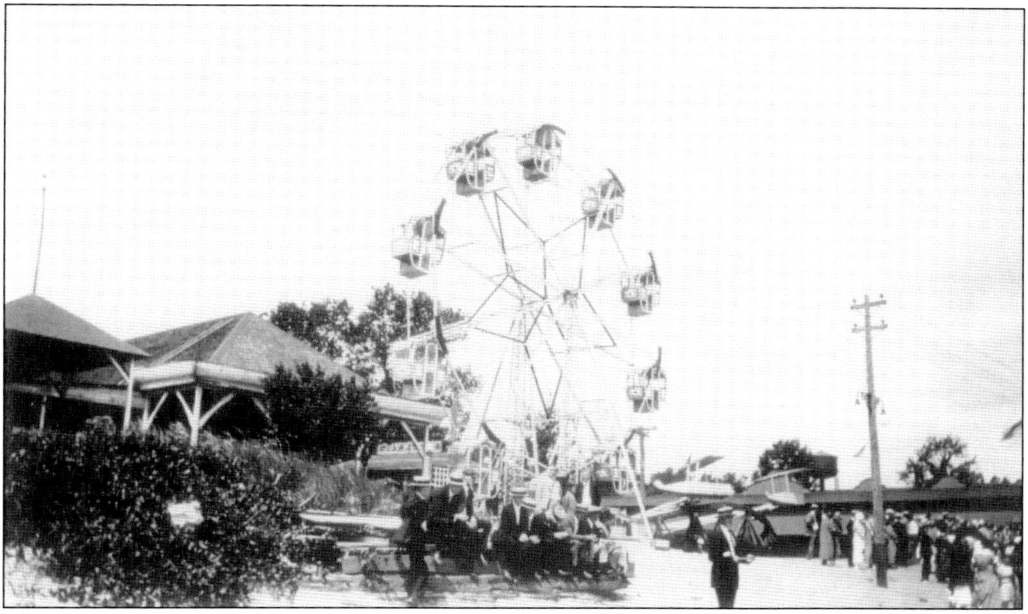

Men in straw hats sit in front of the Ferris wheel at Cedar Point in the 1920s. The water tower in the right background locates this setting as very close to the present site of the 145-foot Giant Wheel at Cedar Point. At the time of this photograph, Cedar Point had an impressive lineup of rides, including several roller coasters. (Courtesy of Ohio History Connection.)

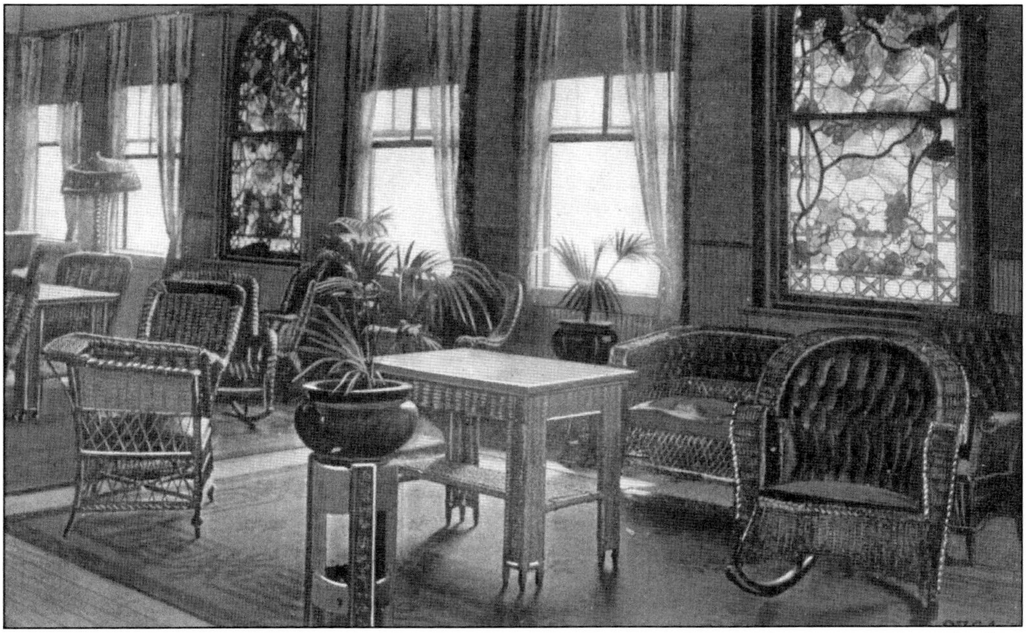

The Cedars Hotel, a less showy and more informal hotel than the Breakers, opened on the bay side of Cedar Point in 1915. It incorporated the buildings that were the original White House Hotel. In addition to 270 guest rooms, the new hotel also featured a cafeteria and gift shop, as well as areas furnished with wicker chairs and writing tables. Today, it is a dormitory for seasonal employees. (Courtesy of Frank Landl.)

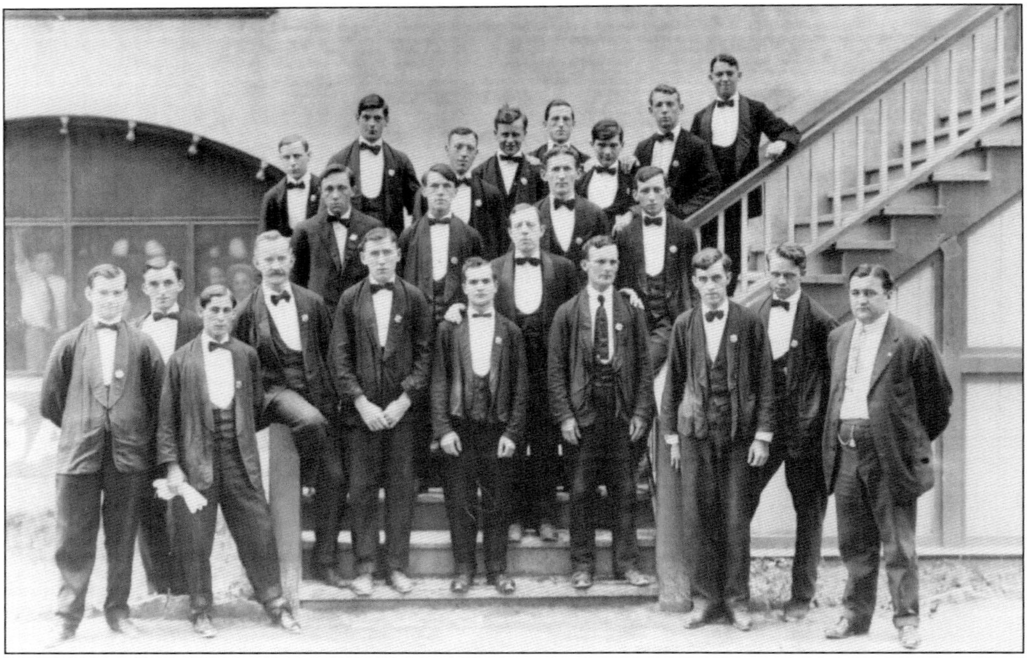

Members of the serving staff at the Rathskellar at the Cedar Point Coliseum pose in 1918. Their formal dress reflects George Boeckling's ambition to make Cedar Point a place that would compare with the best resorts in the United States. Note the young man in the background at left, inside the building, staring at the camera. (Courtesy of HPC.)

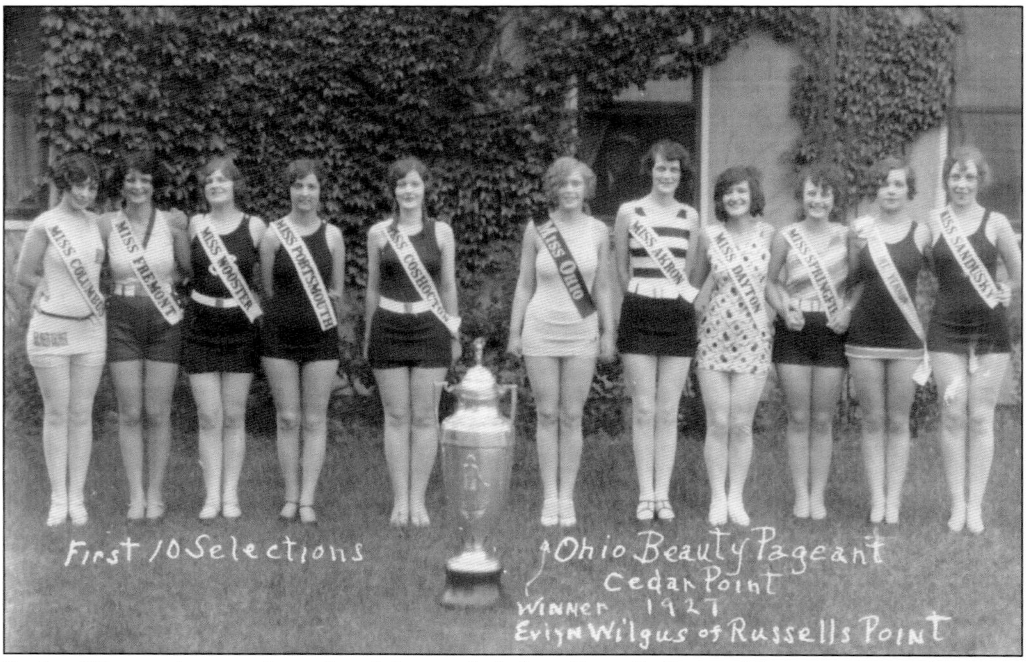

Evelyn Wingus (just to the right of the trophy), from Russells Point, Ohio, was Miss Ohio 1927. The contestants are posing at the Hotel Breakers. Over the years, Cedar Point has hosted many beauty pageants. In the background, an interested observer in a straw hat looks on from a hotel window. (Courtesy of SLARC.)

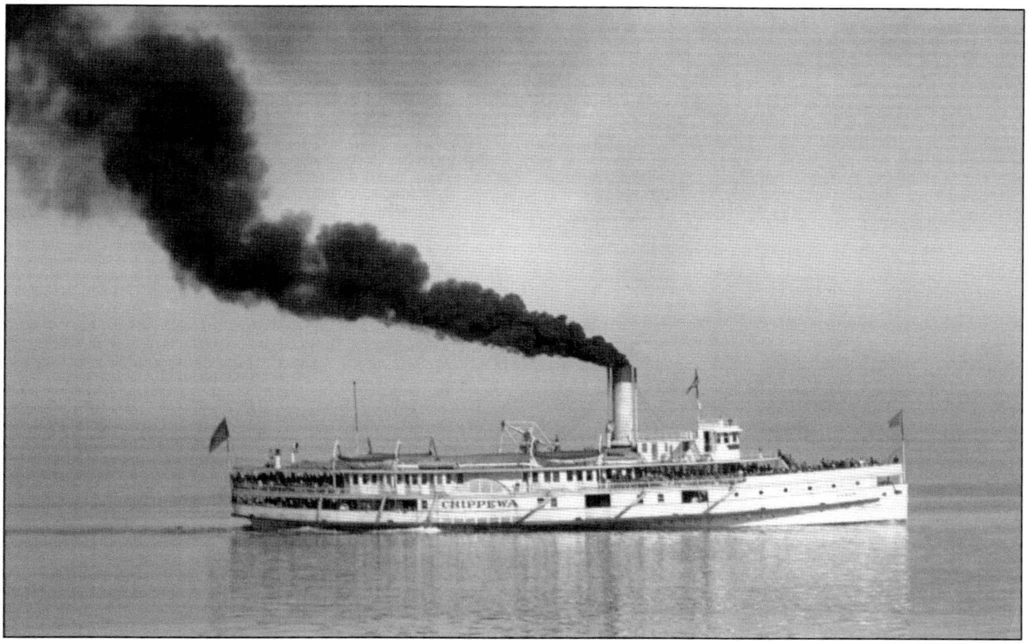

The *Chippewa* steams through western Lake Erie around 1923. The *Chippewa* had an interesting history, starting life as an armed US revenue cutter in 1883. She was converted to a passenger ship in 1922 by the Sandusky and Islands Steamboat Company and operated on the route connecting Sandusky, Kelleys, Lakeside, Middle Bass, North Bass, and Put-in-Bay. In 1938, she made her last Sandusky and Islands run and in 1942 came to the same inglorious end as all steamers: she was cut up for scrap. (Courtesy of SLARC.)

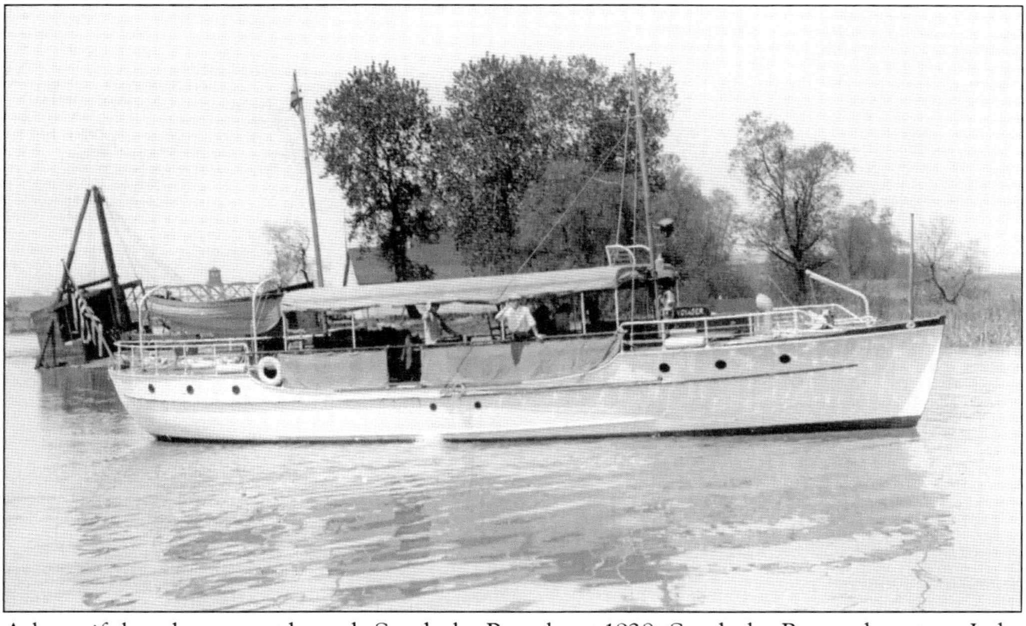

A beautiful yacht moves through Sandusky Bay about 1928. Sandusky Bay and western Lake Erie were, and are, home to an amazing assortment of pleasure boats during the boating season (mid-May to the end of September). In the 1920s, reflecting the booming economy, the number of pleasure craft increased. The Sandusky Yacht Club was founded in 1894. (Courtesy of SAMA.)

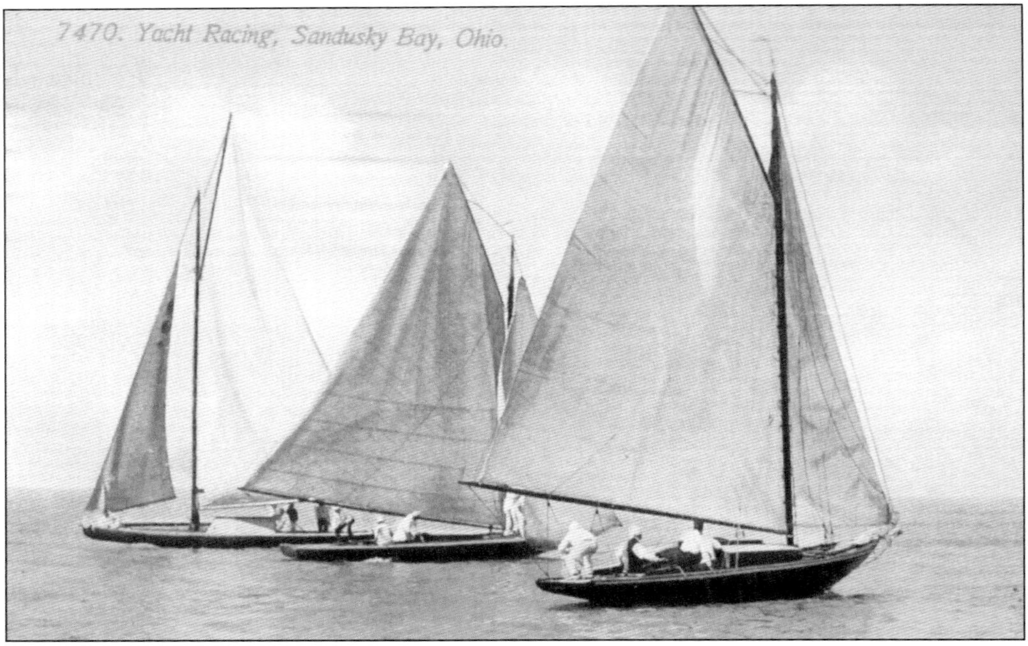

Sailboats race across Sandusky Bay in this postcard from around 1920. The bay, with its large but protected waters, is an ideal location for sailboat racing. The Sandusky Sailing Club, founded in 1932, set up organized sailboat racing in Sandusky Bay. (Courtesy of SAMA.)

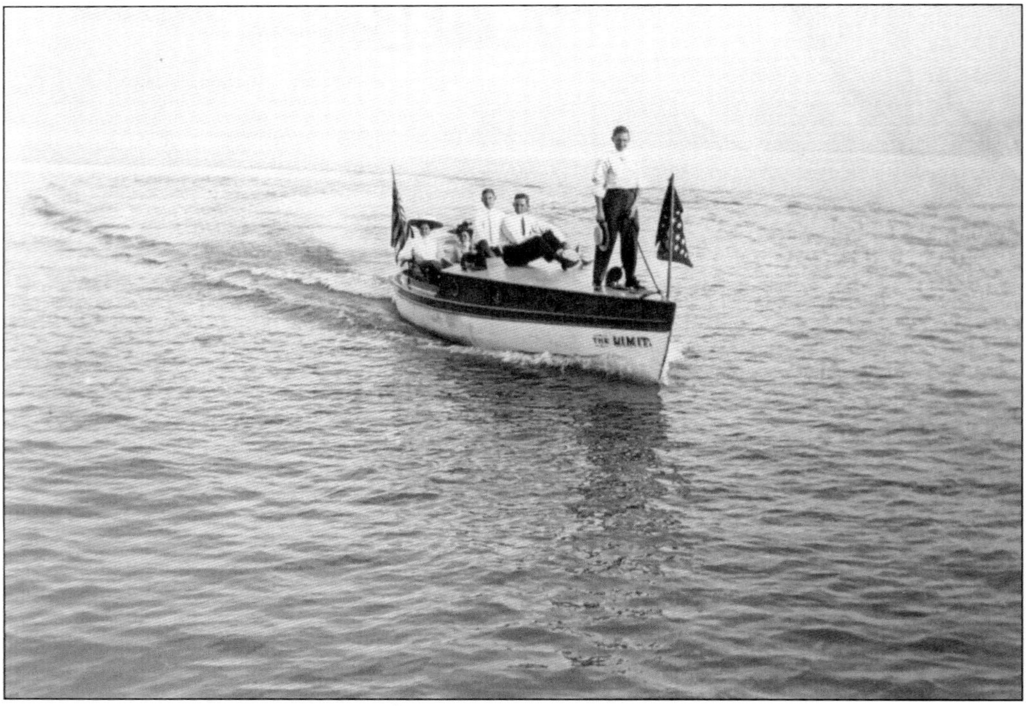

Here, three men and two women motor across Sandusky Bay in the boat *The Limit* sometime between 1920 and 1930. Note that two of the men are wearing ties, and one of the women sports a large hat in the style of the day. Sandusky Bay was, and is, a more protected boating option when Lake Erie has rough water. (Courtesy of SLARC.)

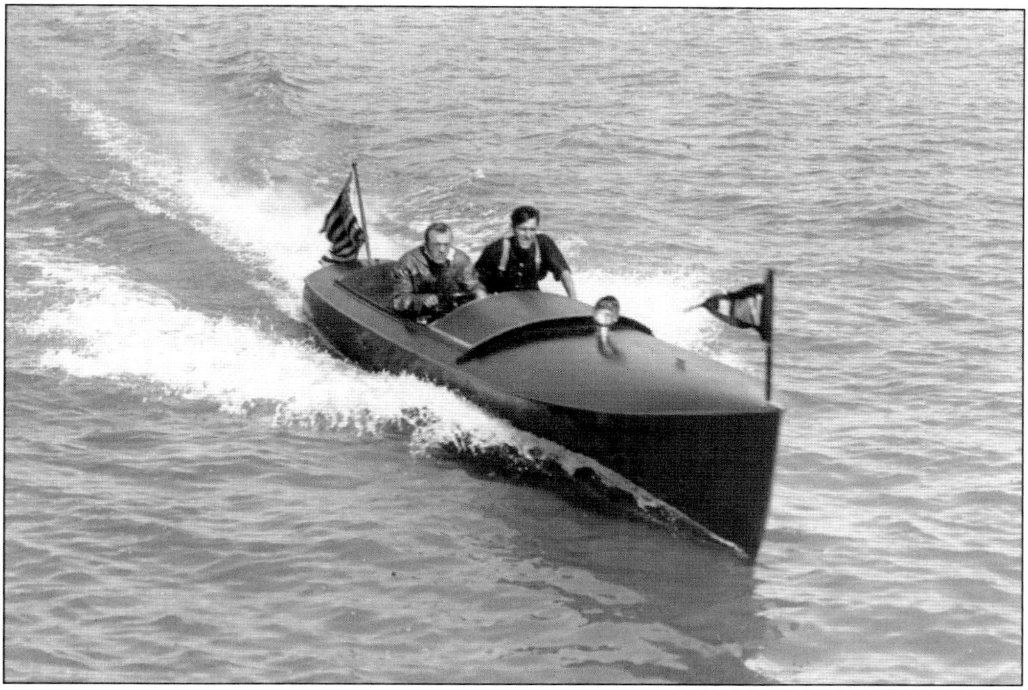

A beautiful wooden motorboat with a VIM flag at the bow knifes through the water on Sandusky Bay between 1925 and 1930. This is a postcard for the VIM Motor Manufacturing Company of Sandusky. (Courtesy of Frank Landl.)

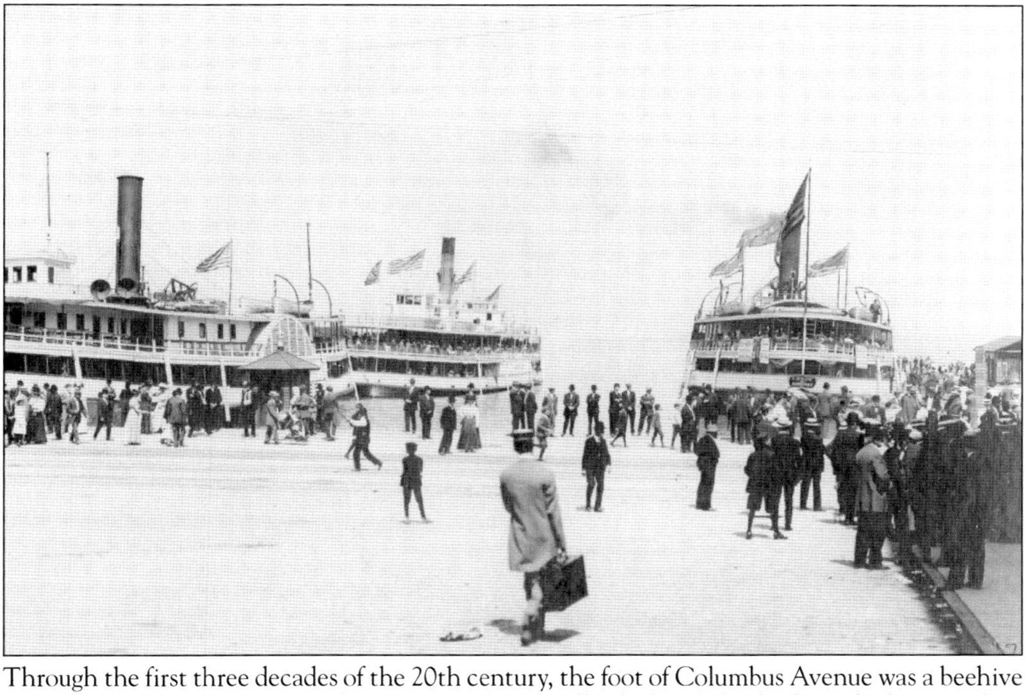

Through the first three decades of the 20th century, the foot of Columbus Avenue was a beehive of activity in the summer months. However, boat traffic declined slowly through the period, as more and more visitors drove automobiles to Cedar Point. This photograph was taken about 1918. (Courtesy of HPC.)

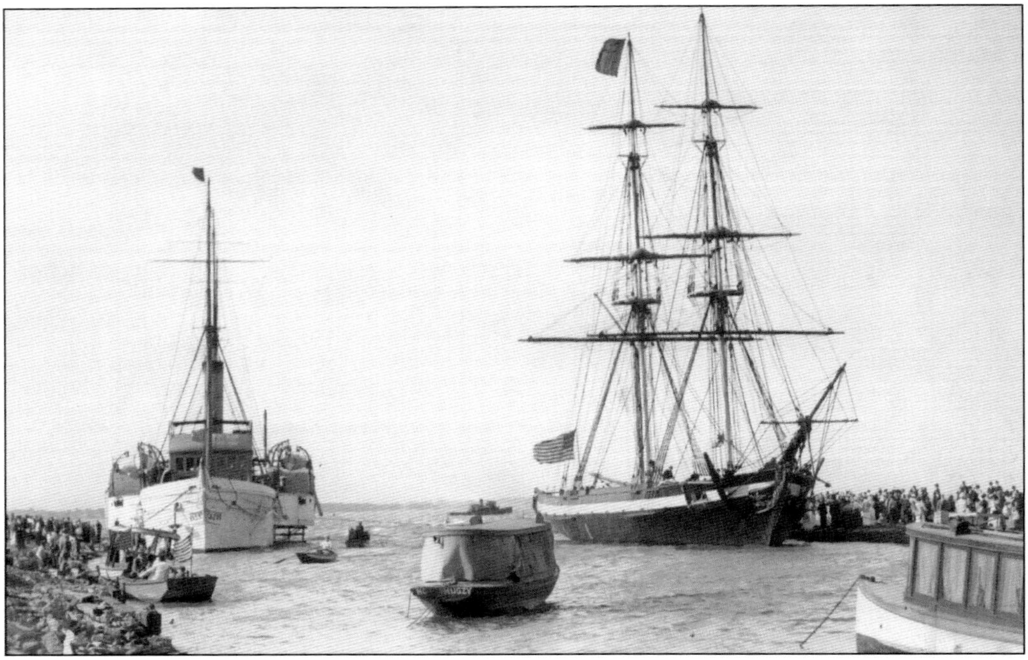

In September 1913, Commodore Oliver Hazard Perry's reconstructed flagship, the *Niagara*, visited Sandusky as part of the centennial celebration of the Battle of Lake Erie. The ship was docked at the Jackson Street pier to the cheers of thousands. The centennial was a major tourist event for the shores and islands area. (Courtesy of SAMA.)

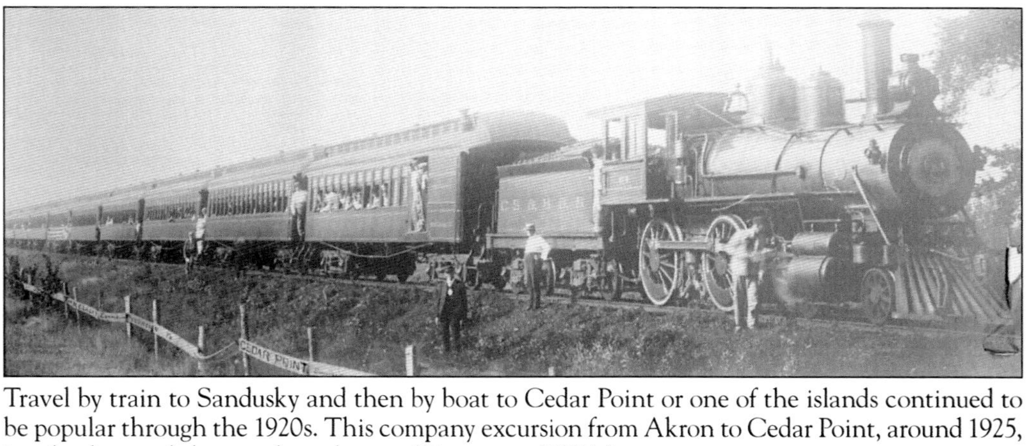

Travel by train to Sandusky and then by boat to Cedar Point or one of the islands continued to be popular through the 1920s. This company excursion from Akron to Cedar Point, around 1925, involved several thousand employees. (Courtesy of CPA.)

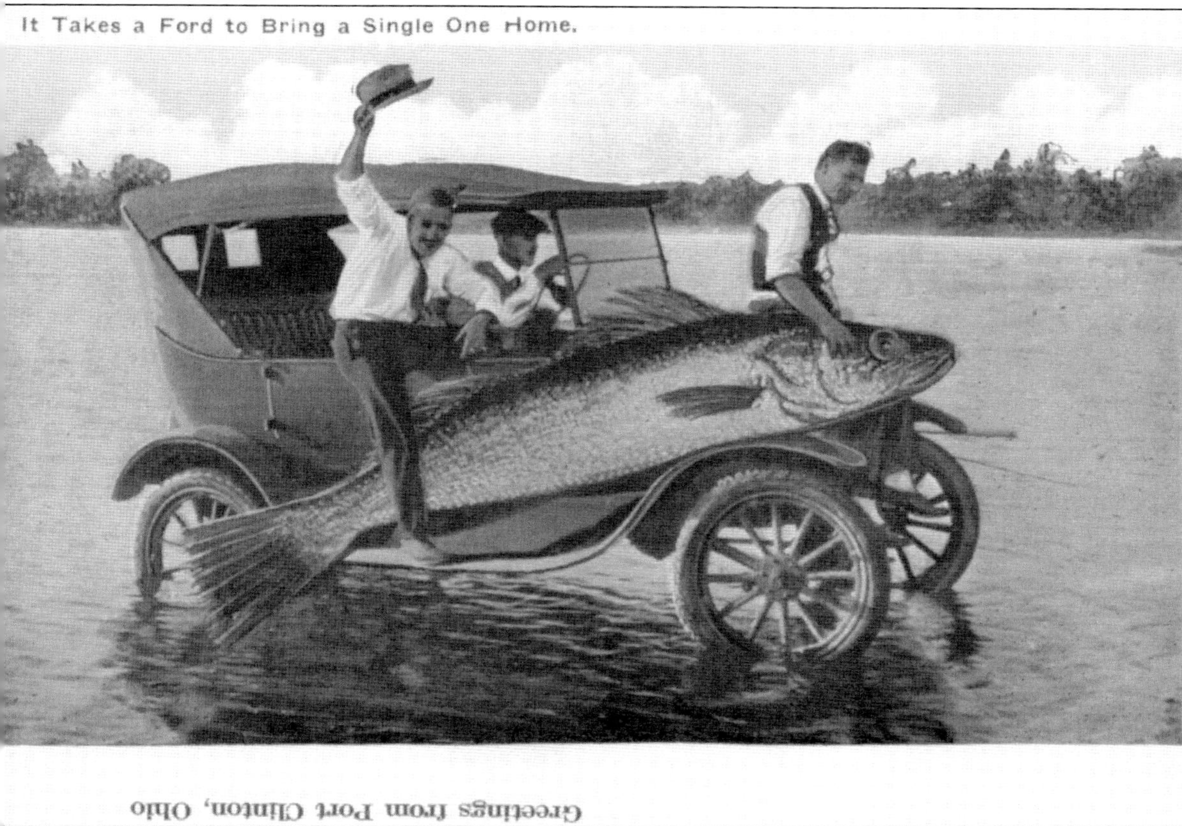

This c. 1930 postcard from Port Clinton suggests there are some very big fish in western Lake Erie, big enough that "it takes a Ford to bring a single one home." Whimsical postcards such as this have a long tradition and are very collectable. (Courtesy of Frank Landl.)

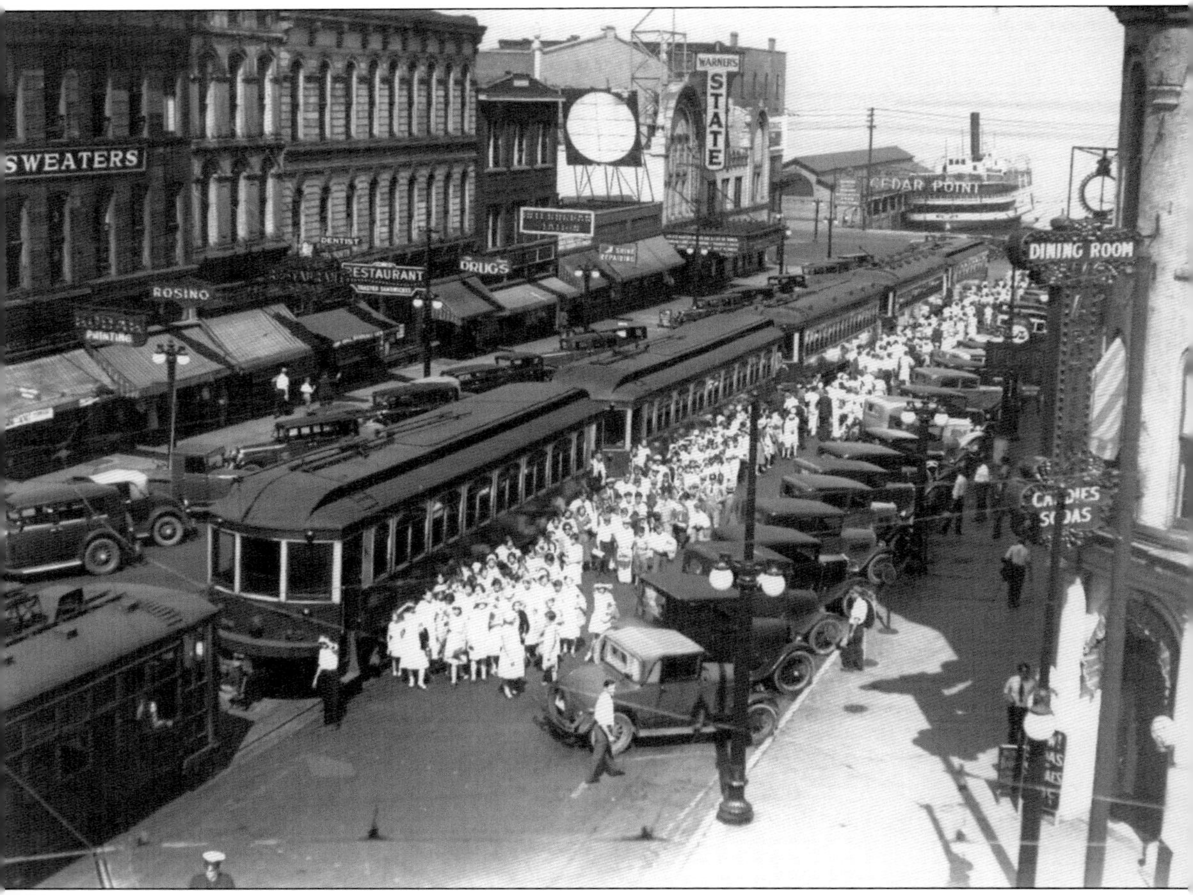

The Lake Shore Electric interurban delivers hundreds of 4H Club girls for a September 1931 outing to Kelleys Island. The steamers G.A. *Boeckling* and *Chippewa* are in the background. The days of the interurbans and lake steamers were soon to come to an end. (Courtesy of HPC.)

Four

Surviving the Depression and the War Years 1932–1949

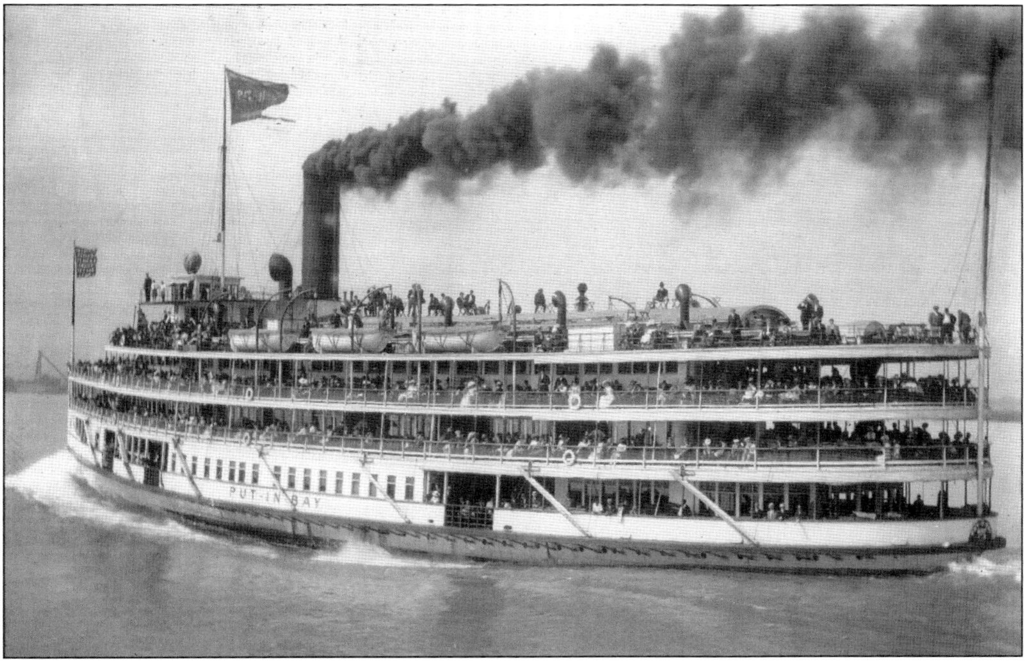

The four-deck steamer *Put-in-Bay*, perhaps the most popular and beloved of all the lake steamers, was built in Detroit in 1910 and operated in western Lake Erie for more than 40 years, retiring in 1949. Here, she is headed down the Detroit River, bound for Lake Erie in the 1920s. The *Put-in-Bay* could transport more than 3,000 passengers and offered dancing to an orchestra in its main salon. The tough economy of the 1930s and the growth of the automobile in the 1940s brought the era of the lake steamers to an end. (Courtesy of SLARC.)

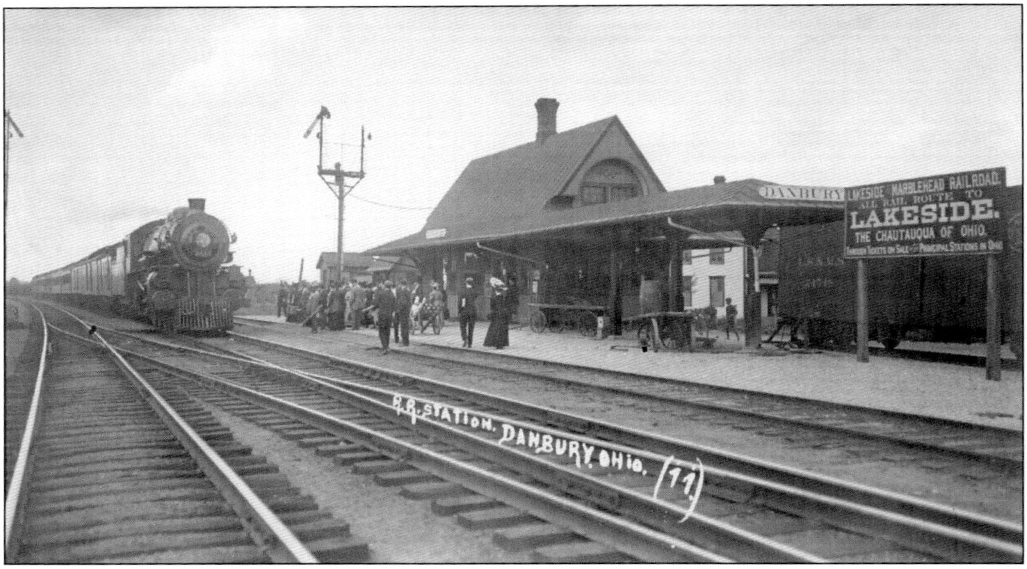

Train travel continued to bring visitors to Lake Erie communities in the 1930s and 1940s. The Lakeside & Marblehead Railroad carried visitors to Lakeside, the "Chautauqua of Ohio." The town's depot is seen here around 1949. Lakeside was able to maintain its visitation and programming through this period. First Lady Eleanor Roosevelt was a speaker and visitor in July 1940. (Courtesy of HPC.)

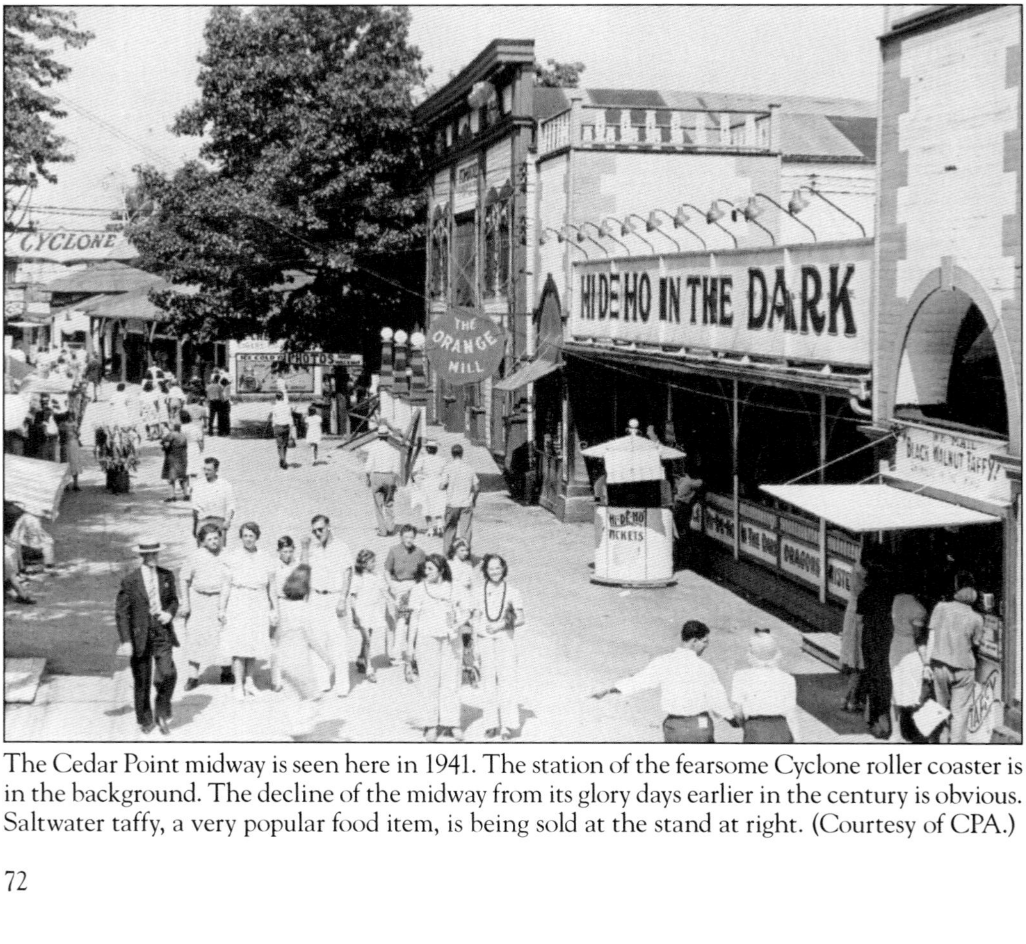

The Cedar Point midway is seen here in 1941. The station of the fearsome Cyclone roller coaster is in the background. The decline of the midway from its glory days earlier in the century is obvious. Saltwater taffy, a very popular food item, is being sold at the stand at right. (Courtesy of CPA.)

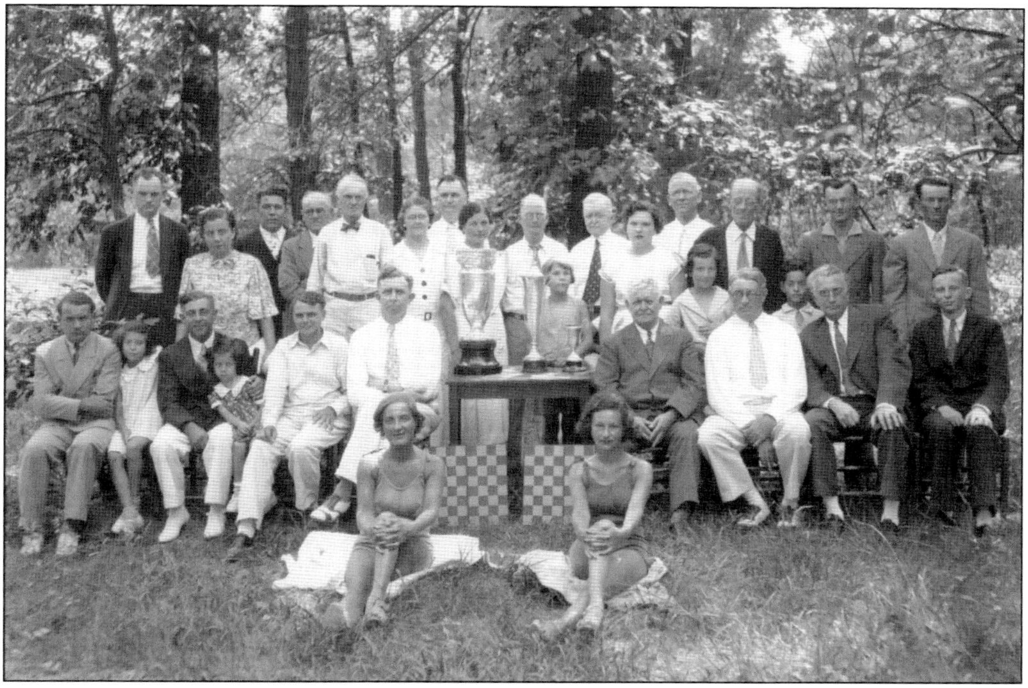

In the summer of 1936, Cedar Point was the host site for a convention and competition of the nation's premier checkers players. Note the rather formal attire for a summer resort, as well as the serious faces (with one exception, probably the winner). (Courtesy of HPC.)

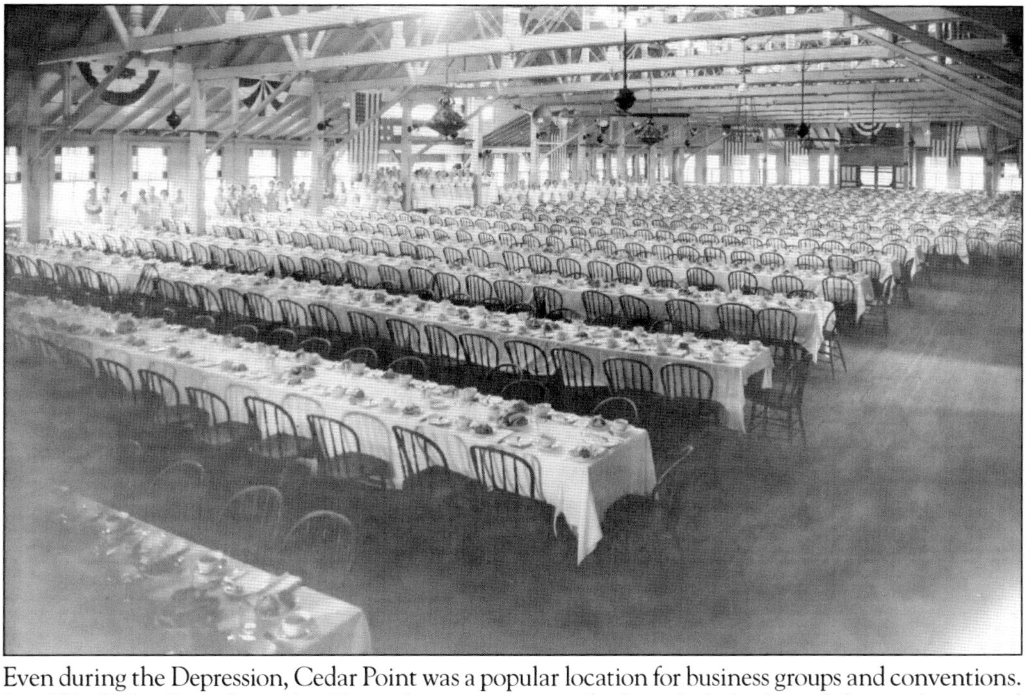

Even during the Depression, Cedar Point was a popular location for business groups and conventions. In 1933, Cedar Point hosted a Chevrolet convention, which included a formal dinner for several hundred conventioneers. The location of the dinner is the second floor of the original Grand Pavilion. (Courtesy of HPC.)

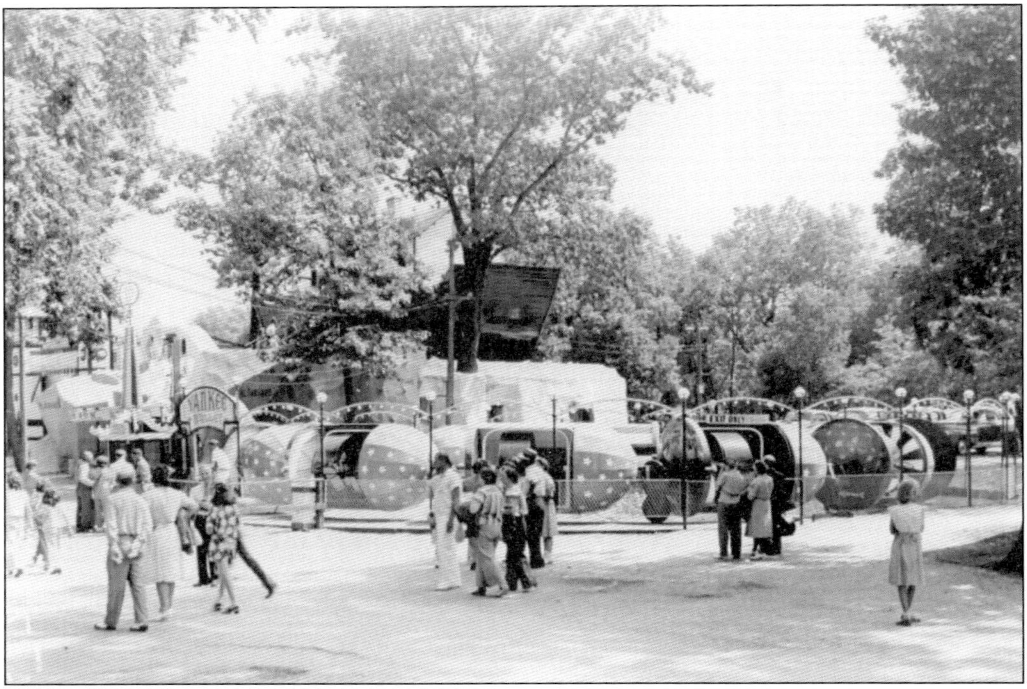

The Yankee Bullet was a popular ride on the Cedar Point midway in the 1940s. In this 1941 photograph, Noah's Ark, also a popular attraction, is the background. Investment in new rides and attractions was minimal in the 1930s and 1940s. (Courtesy of CPA.)

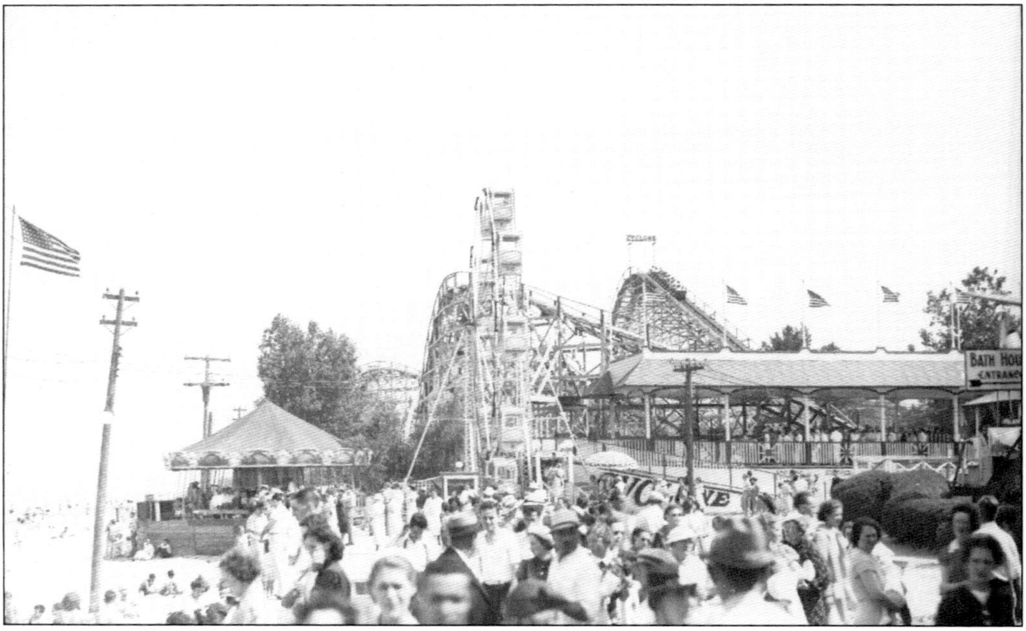

The Beach midway is seen here around 1940, with the Cyclone and Ferris wheel in the background. Designed by the legendary roller coaster builder Harry Traver, the Cyclone was one of the best wooden coasters ever constructed in the Midwest. It opened in 1929 and operated for a quarter of a century. It was considered a terrifying ride. Note the large number of American flags in the photograph. (Courtesy of CPA.)

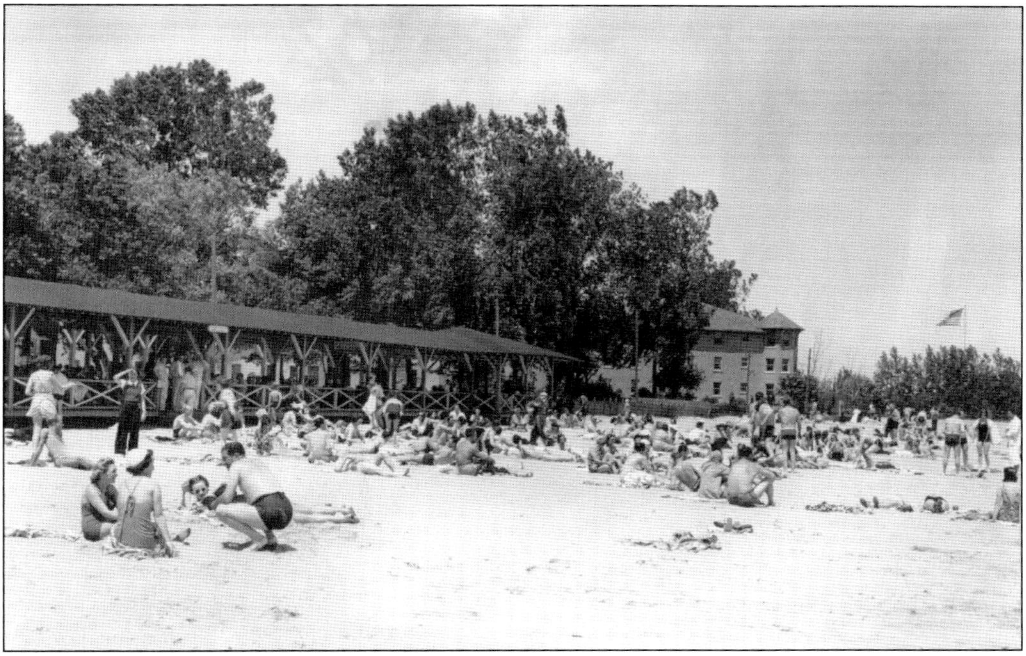

The Cedar Point Beach is seen here around 1940, with the Hotel Breakers in the background. Cedar Point's bathing beach remained popular with guests, a constant in all eras. Note the large American flag on the right. Cedar Point's tradition of prominently flying the American flag has also been a constant over the years. (Courtesy of SLARC.)

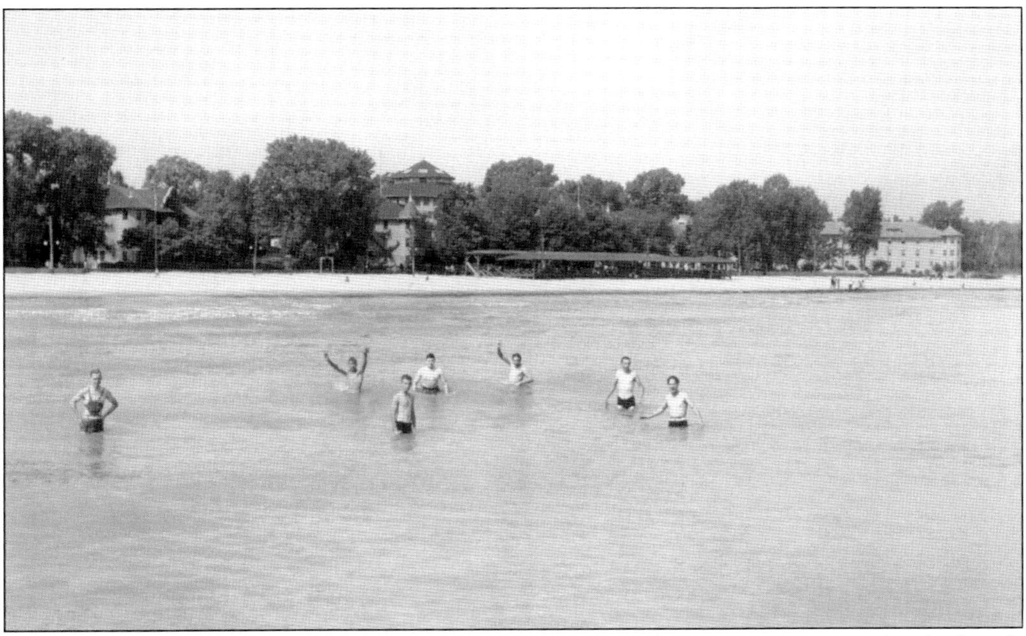

Bathers mug for the camera off the Cedar Point Beach in the 1940s. Interestingly, the beach behind them is nearly empty. The photograph was likely taken early in the day. Unlike most beach photographs, the perspective here is from the water to the land. The photographer, likely positioned on scaffolding, was aided by the fact that the water remained shallow well out into the lake. (Courtesy of CPA.)

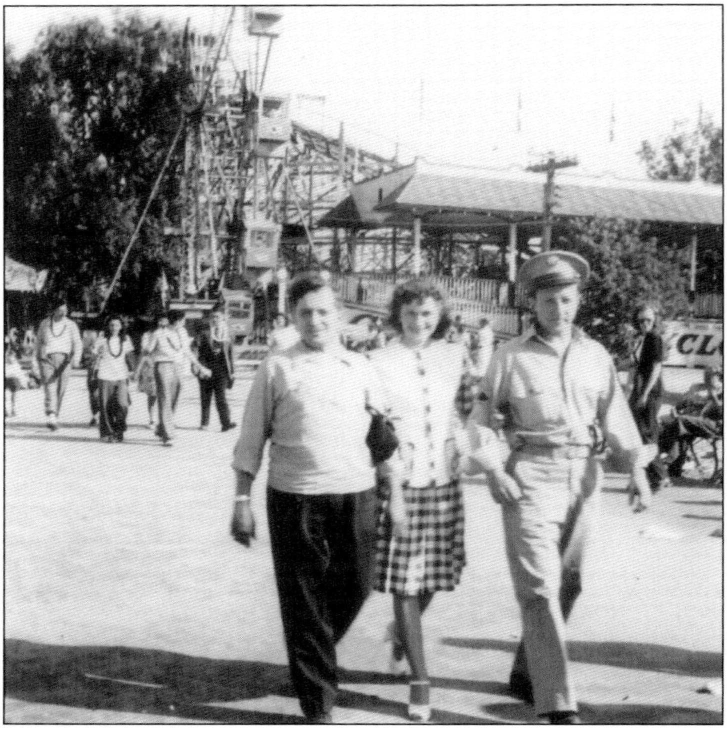

During World War II, Cedar Point was a popular spot for soldiers on leave. This soldier, Hugo Hildebrandt of Cleveland, has just ridden the Cyclone (the station is in the background). With him are his brother Joe and his brother's wife, Coletta Stafford. The soldier's wife, Ruth, took the photograph, dated to 1942–1943. (Courtesy of the Hildebrandt collection.)

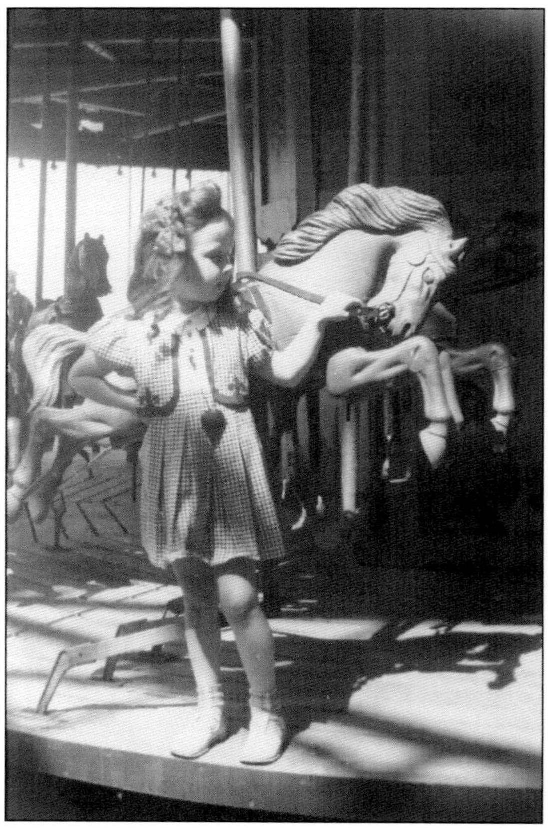

In 1945, adorable Sheila Stanley, five years of age, strikes a pose on the children's carousel at Cedar Point, which was located on the beach nearly adjacent to the famous Cyclone roller coaster. Cedar Point's carousel history has included many famous machines, including the Midway Carousel, created by master carver Daniel Muller, which debuted at Cedar Point in 1946. (Courtesy of Sheila Ehrhardt.)

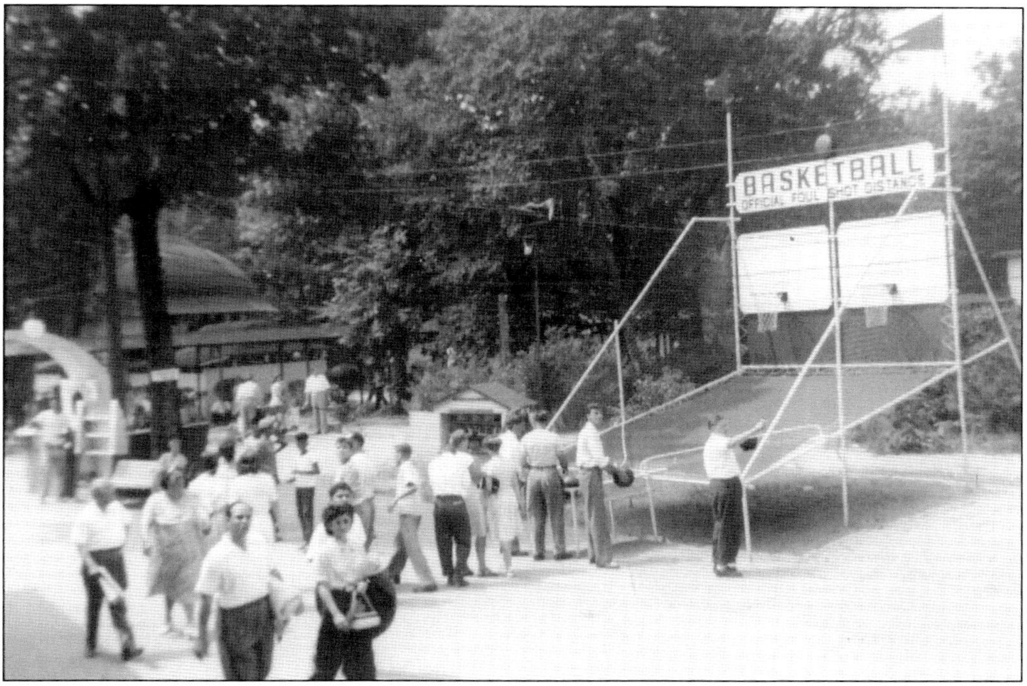

Seen here around 1947 is a basketball game on the Cedar Point Midway. Nearly 70 years later, basketball games are still among the most popular and profitable in the park. Midway games have been an important part of the amusement park experience since Boeckling's era. In the 1930s and 1940s, most games were operated by concessionaires. (Courtesy of Sheila Ehrhardt.)

One bright spot for Cedar Point during the hard times of the 1930s and World War II was the conversion of the second floor of the Coliseum into a ballroom. All of the big-band musicians of the time, from Tommy Dorsey to Guy Lombardo, played to large crowds in the Cedar Point Ballroom, an Art Deco masterpiece that still looks the part. These dancers are seen around 1939. (Courtesy of CPA.)

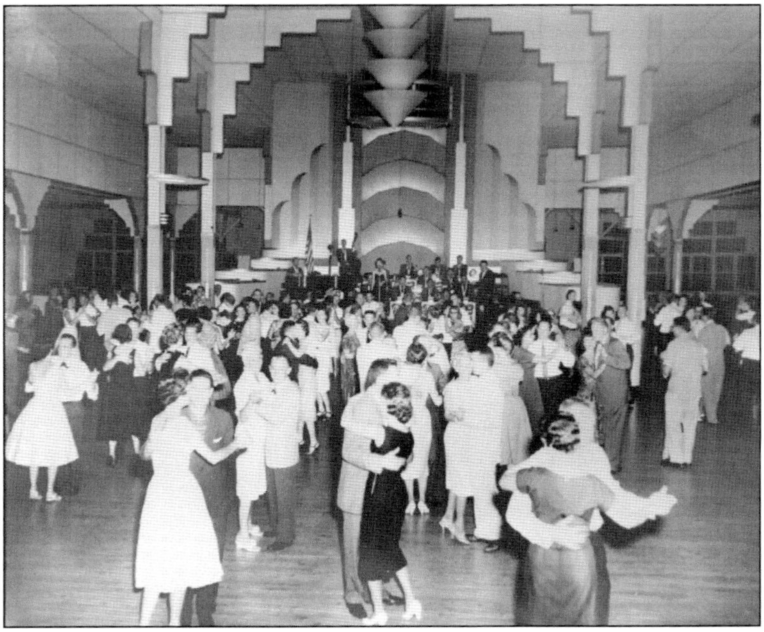

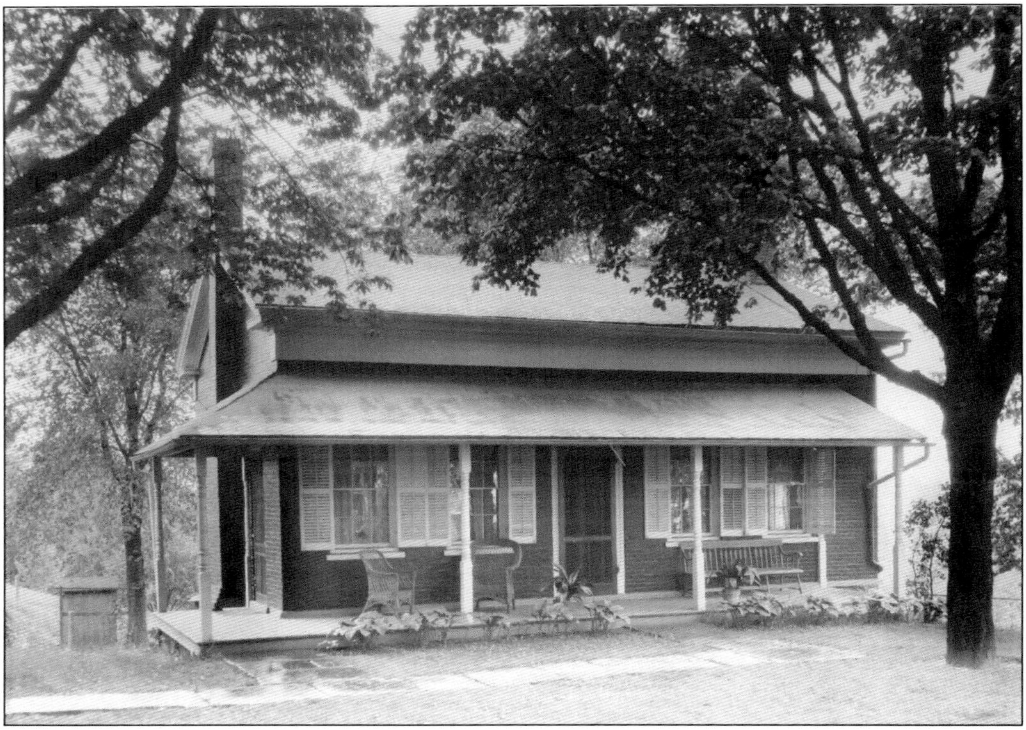

America's most famous inventor, Thomas A. Edison, was born in the village of Milan, just south of Sandusky, in 1847. The Edison Birthplace Museum opened on the centennial of Edison's birth in 1947. Milan also is home to the Milan Museum, a six-building complex that features a well-known glass collection. (Courtesy of HPC.)

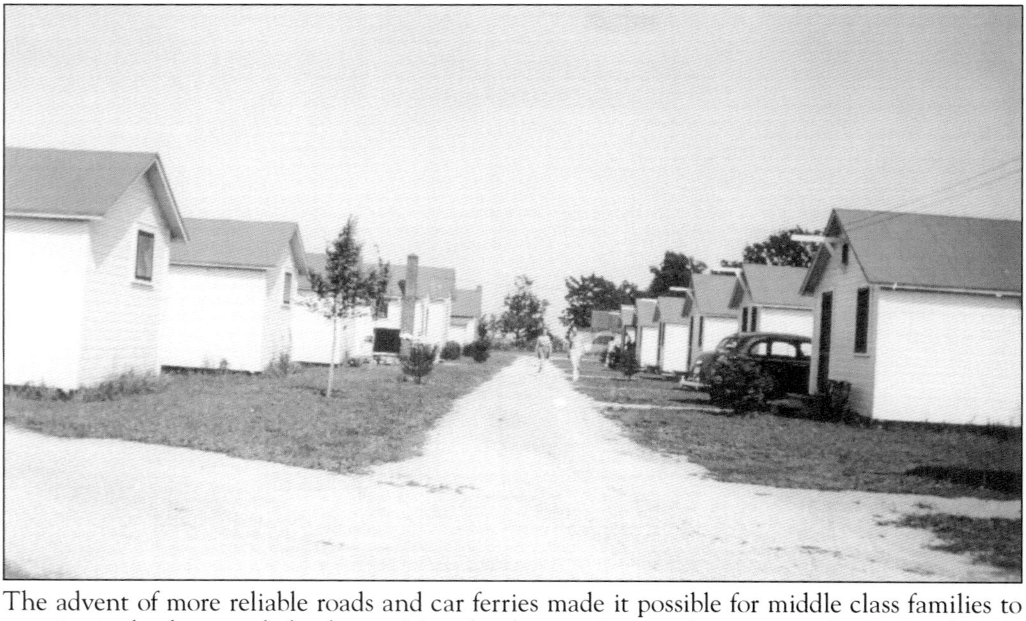

The advent of more reliable roads and car ferries made it possible for middle class families to vacation in the shores and islands area. Many families were looking for economical accommodations. These cabins are typical of many cabin developments on the Lake Erie islands and shoreline. This photograph dates to the 1930s or 1940s. (Courtesy of Ohio History Connection.)

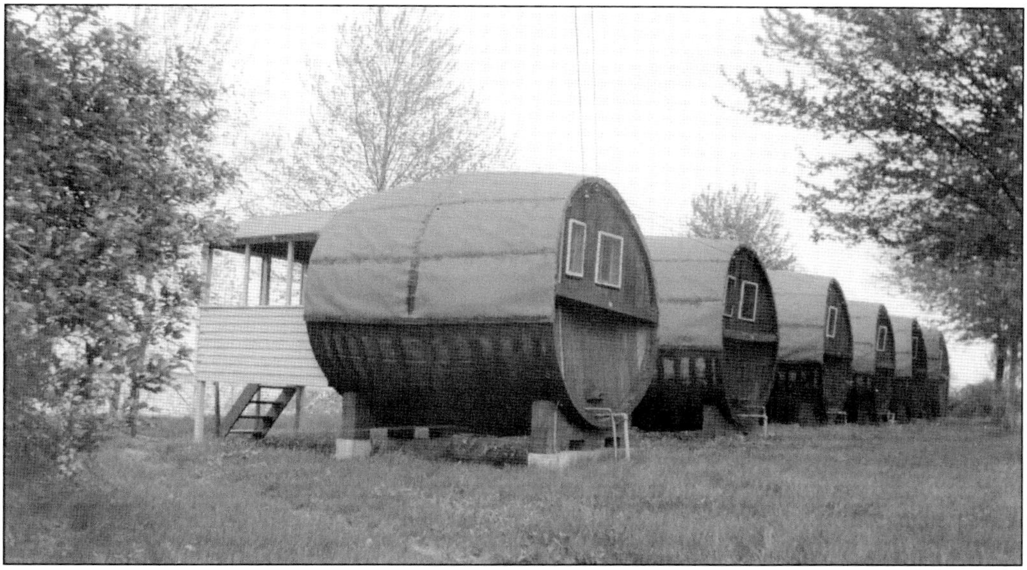

When it came to vacation cottages, builders often tried to gain a competitive edge with more visual appeal. Vacation rentals designed to look like wine casks, seen here in the 1930s or 1940s, were available on the shoreline between Vermilion and Huron and also on South Bass Island. (Courtesy of Ohio History Connection.)

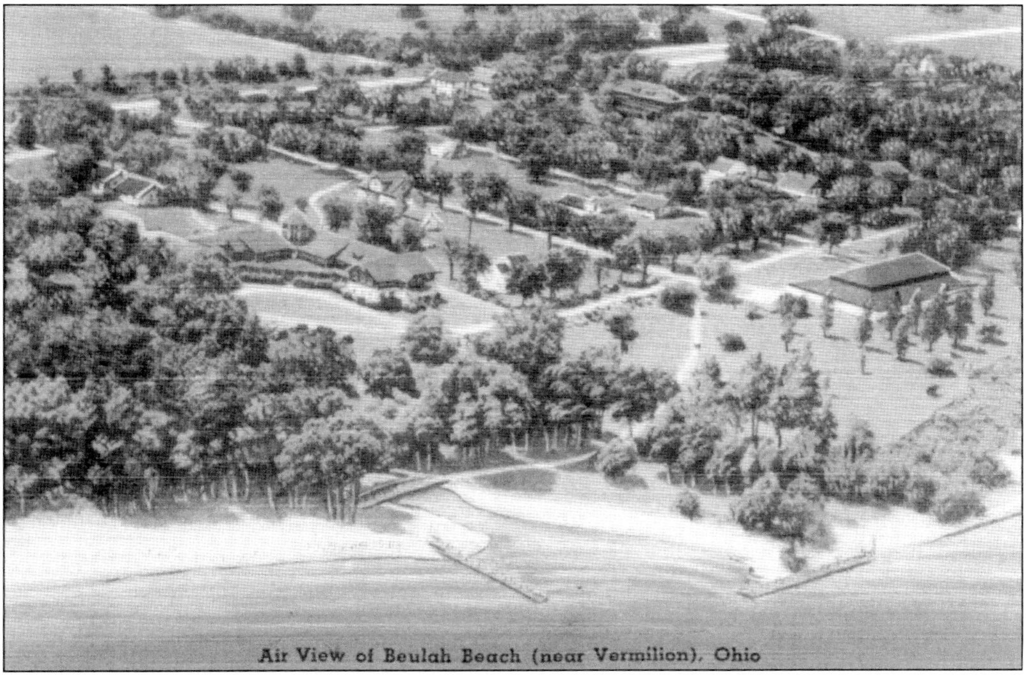

Beulah Beach was one of several religious-based communities west of Vermilion. All of the communities benefited from beautiful beaches, attractive countryside, and good transportation (Lake Shore Electric Railway and US Route 6). In this 1949 postcard, the 132-acre parkland of the Central District of the Christian Missionary Alliance shows the tabernacle (right), the institute (left), the missionary (upper right), and the beach. (Courtesy of Frank Landl.)

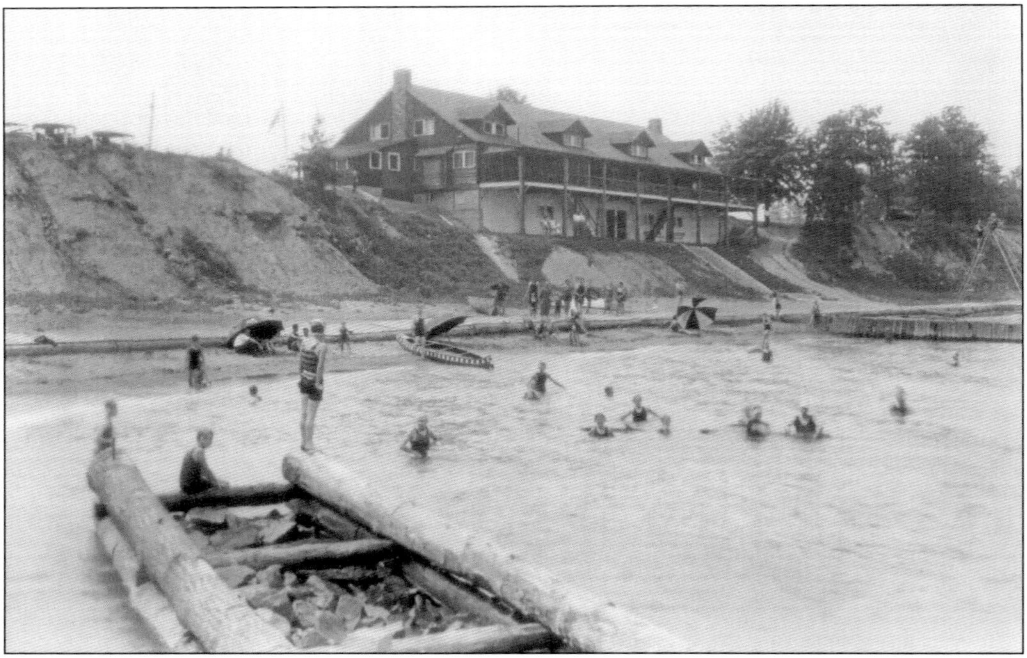

Founded by a group of Cleveland investors in 1919, the vacation community of Vermilion-on-the-Lake offered cottages, boat and canoe rentals, a beautiful beach, and a large dining and community center called the Vermilion-on-the-Lake Clubhouse (still in use). This photograph dates to the 1930s. (Courtesy of Vermilion Archival Society.)

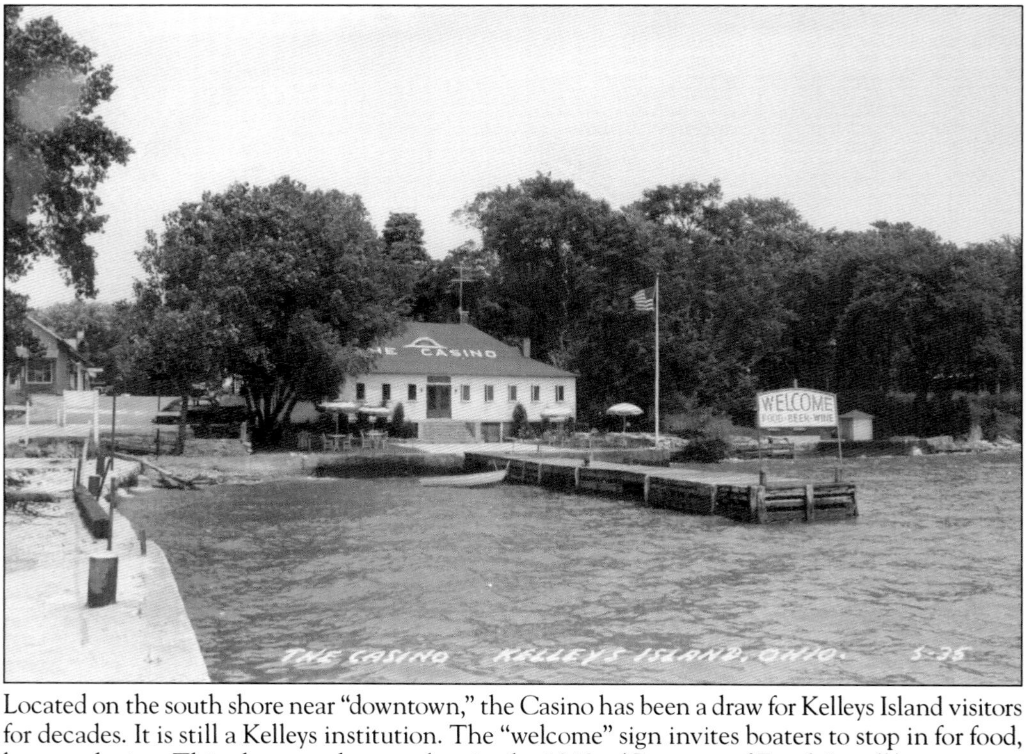

Located on the south shore near "downtown," the Casino has been a draw for Kelleys Island visitors for decades. It is still a Kelleys institution. The "welcome" sign invites boaters to stop in for food, beer, and wine. This photograph was taken in the 1940s. (Courtesy of Frank Landl.)

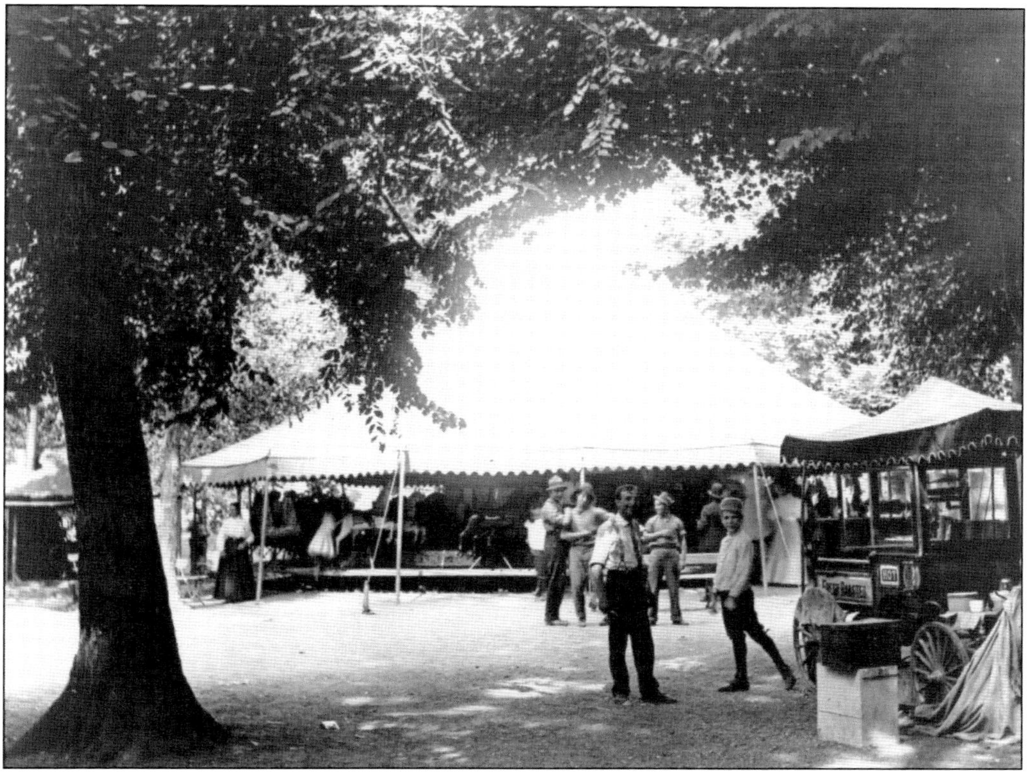

A small group gathers around the carousel at Crystal Beach Park in the 1920s or 1930s. Like most tourist businesses in the region, Crystal Beach suffered financially during the Depression and the war years. Like Cedar Point, Crystal Beach had a large ballroom that hosted the big bands of the Swing Era, including Count Basie and Gene Krupa. (Courtesy of the Merry-Go-Round Museum.)

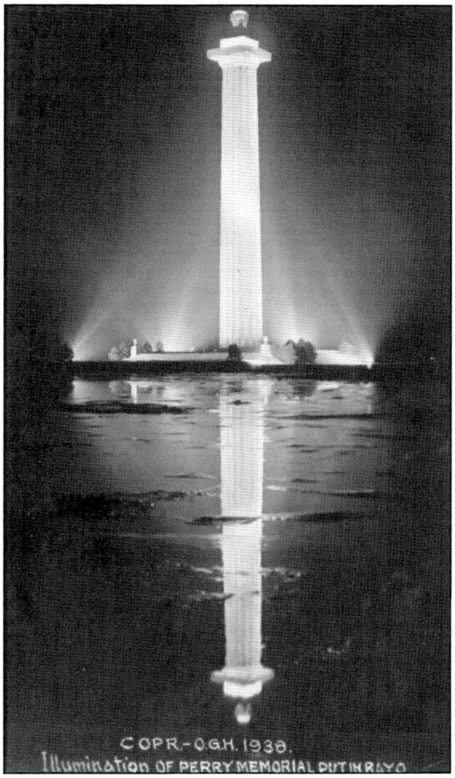

Perry's Memorial is beautifully lit on a calm summer evening in 1938. In addition to its status as a national monument, the structure has served as an important, albeit unofficial, aid to navigation for both pleasure boats and commercial craft in western Lake Erie. The monument can be seen from the beach at Cedar Point on a clear day. (Courtesy of SAMA.)

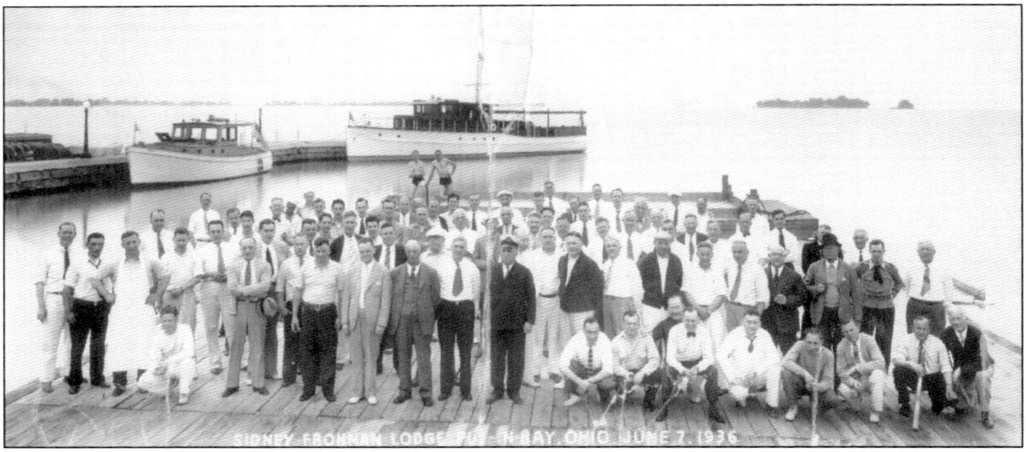

These men are all employees of Sandusky's Hinde & Daugh Paper Company. The 1936 event is an outing at the Put-in-Bay lodge of Sydney Frohman, owner of the company. Most men are in business dress, but note the two men in swimwear in the back. The island in the background is likely Ballast. (Courtesy of SLARC.)

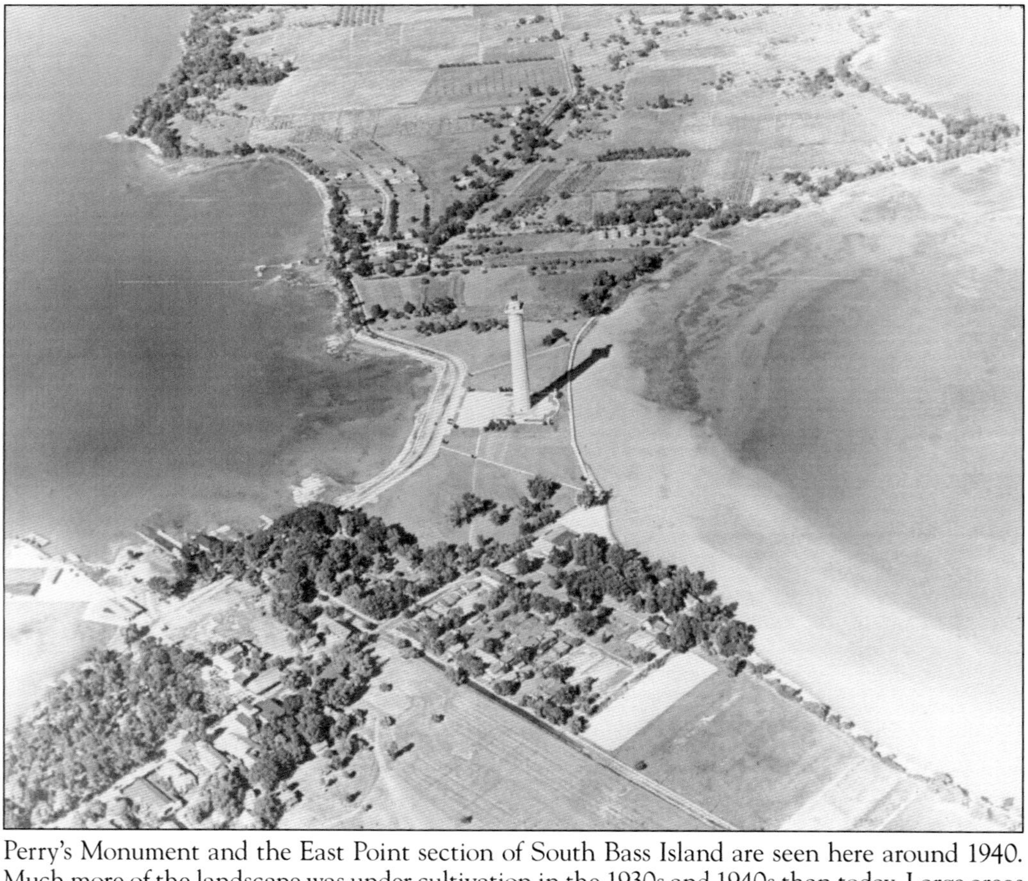

Perry's Monument and the East Point section of South Bass Island are seen here around 1940. Much more of the landscape was under cultivation in the 1930s and 1940s than today. Large areas of the island have since reverted to forest. (Courtesy of Ohio History Connection.)

Ready for a swim are, from left to right, Joe, Art, Margaret, and Hugo Hildebrandt. They stand on an old dock near Peach Point on South Bass Island around 1936. Middle Bass Island is in the background. As the shallowest of the Great Lakes, Erie warms quickly in the summer months, with water temperatures in the 70s. It is ideal for swimming. (Courtesy of James Hildebrandt.)

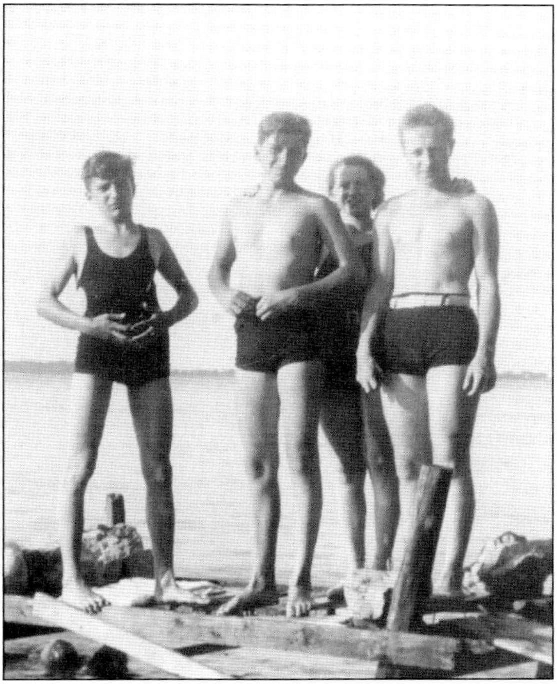

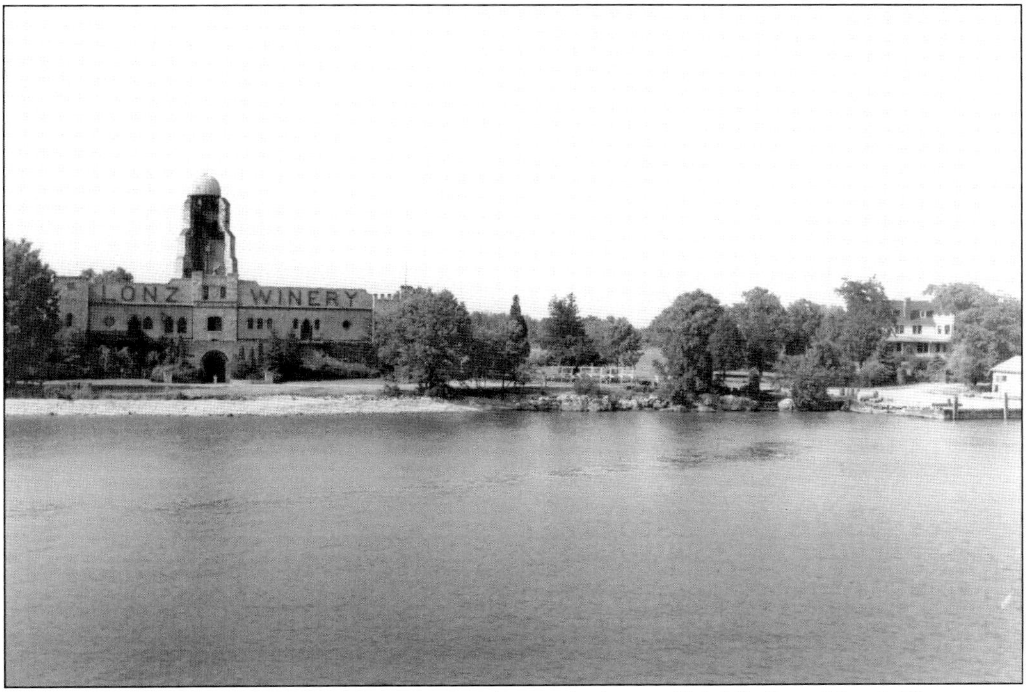

Lonz Winery, seen here in the 1940s, sits on the south shore of Middle Bass Island. The winery was built atop the cellar made by Andrew Wehrle for the Golden Eagle Winery in 1884. George Lonz purchased the winery in 1934, in the aftermath of Prohibition, and designed and built the Moorish castle-like structure, which has become an emblem of the Lake Erie islands. Although the Lonz Winery brand survives, the building and grounds were acquired in 2000 by the State of Ohio for a state park. (Courtesy of James Lang.)

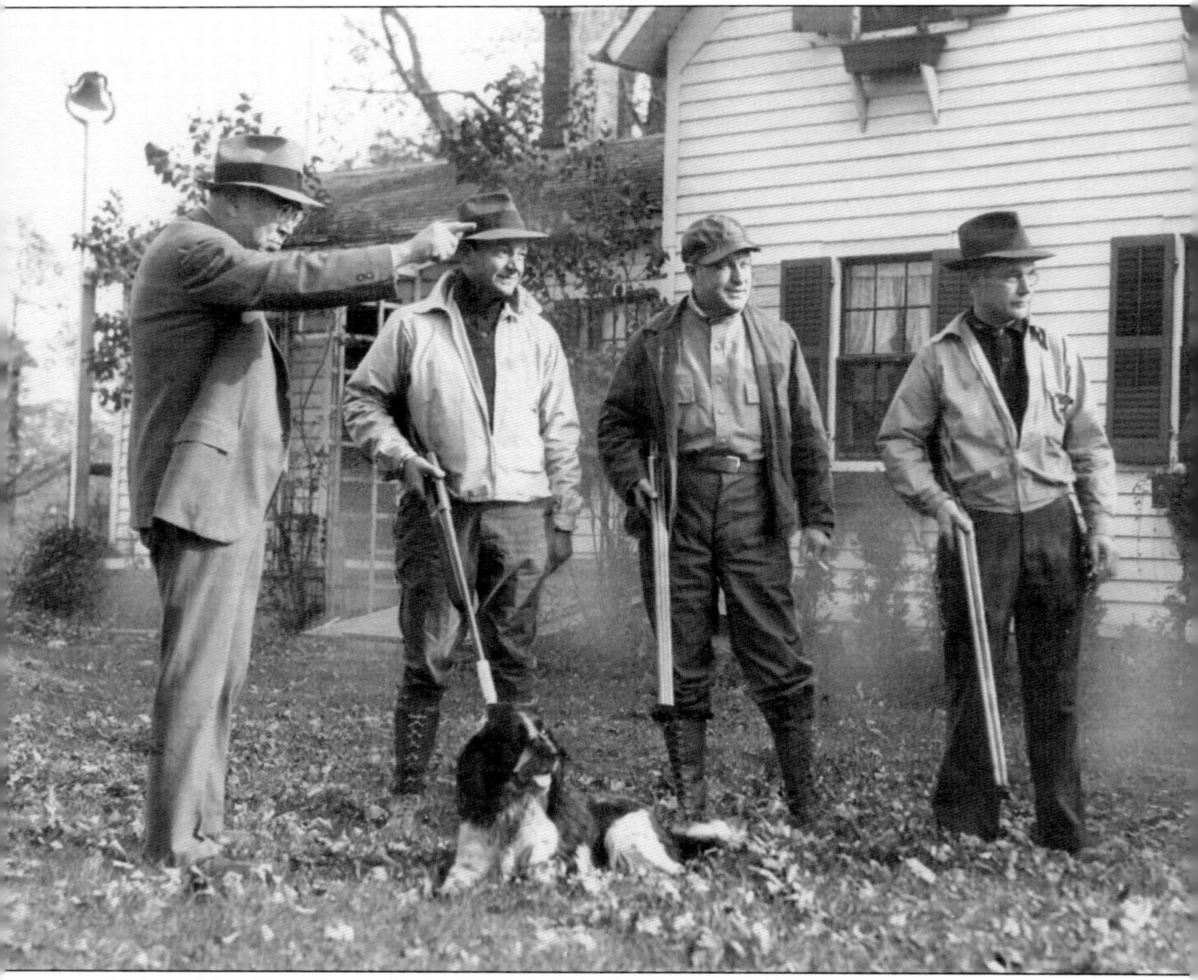

Kelleys Island has always drawn sportsmen for hunting as well as fishing. Waterfowl hunting in particular has been popular. Pheasant hunting was also a draw to the island for many years. In this 1941 *Cleveland Press* photograph, William Minshall (left) of Cleveland and three friends pose with their dog before a pheasant hunt. There are no pheasants on the island today. (Courtesy of Michael Schwartz Library.)

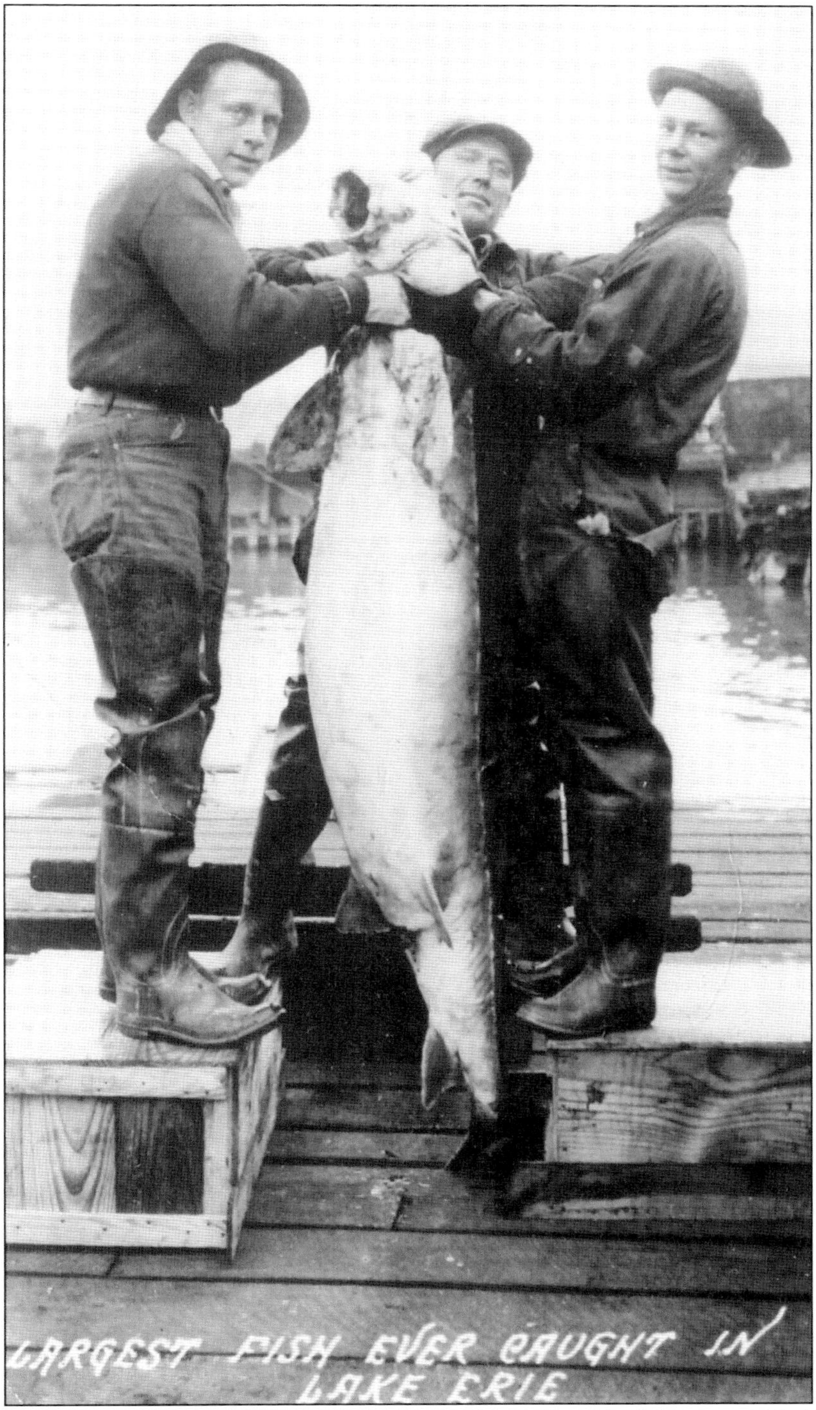

A 180-pound sturgeon caught on April 29, 1935, is proudly displayed by fishermen Alfred McKillips (left), Albert Kuegler (center), and Sylvester Dwelle. Once a common sight in the waters of western Lake Erie, sturgeon are rarely caught today (although some scientists believe the fish is making a subtle comeback). Sturgeon are the largest fish native to the Great Lakes, with some specimens reaching nine feet in length and weighing more than 300 pounds. (Courtesy of HPC.)

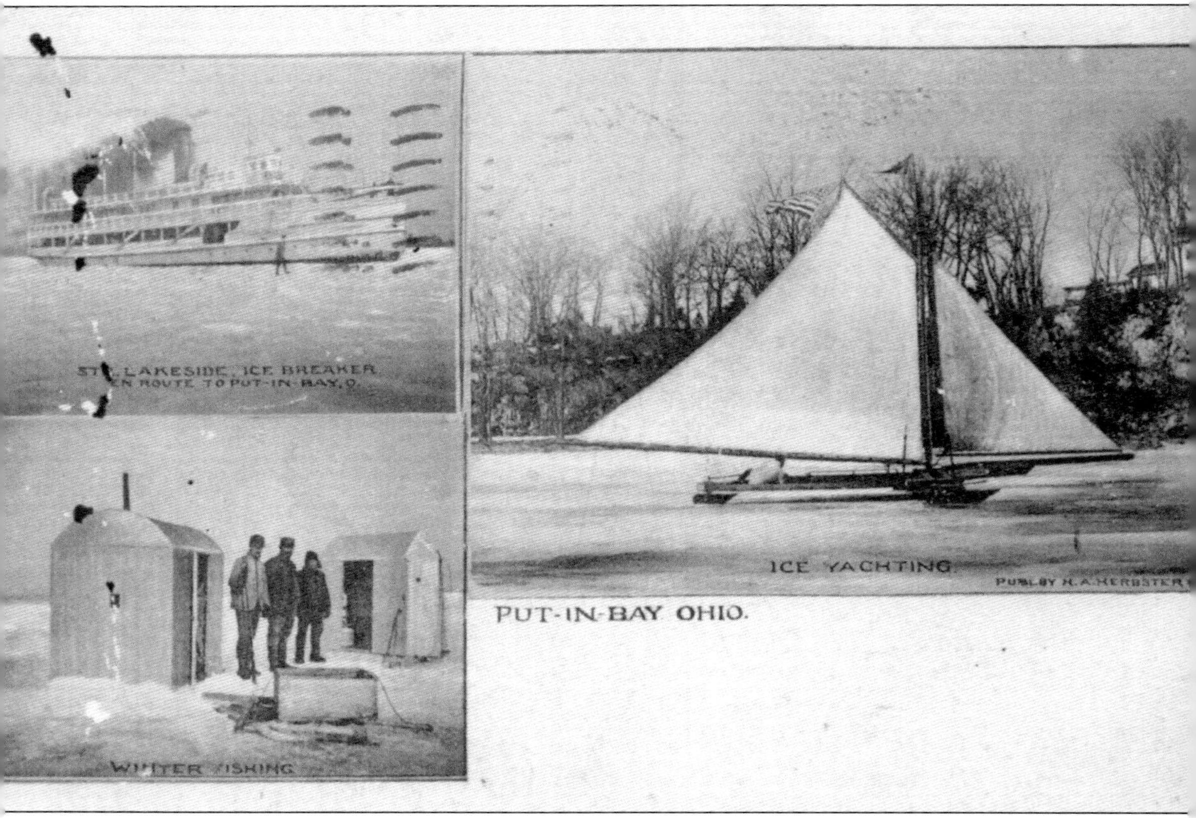

In the 1930s and 1940s, the Lake Erie islands and nearby shore communities like Sandusky continued to promote winter sports, such as ice boating and ice fishing. Put-in-Bay, with its sheltered harbor, was a prime location for both activities. Fishing from an ice shanty or gliding across the ice (sometimes at very high speeds) were good reasons to be outdoors in the winter months. (Courtesy of SAMA.)

Five

Postwar Recovery and the Baby Boom 1950–1969

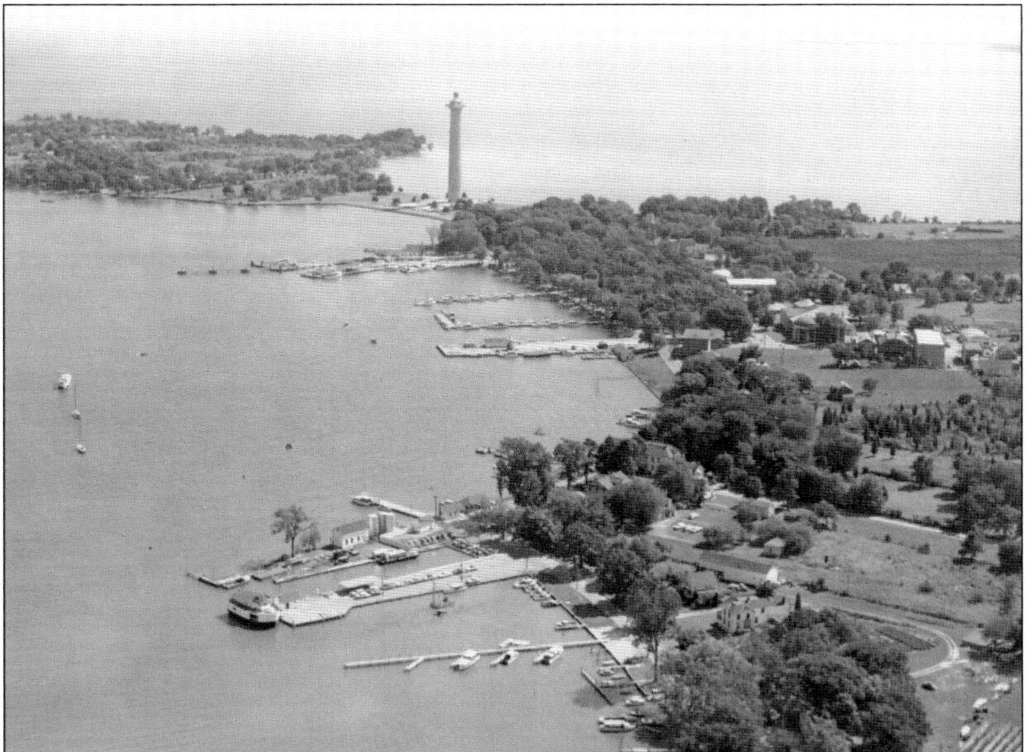

In the aftermath of World War II, the economy roared back to life in the United States and in the Lake Erie shores and islands region. Baby boom families were looking for recreation, and the area's marinas, restaurants, bait shops, motels, campgrounds, wineries, and parks saw increased visitation in the 1950s. This 1950s Tom Root aerial photograph shows South Bass Island. The east-facing photograph includes the public docks at Put-in-Bay and Perry's Monument. (Courtesy of HPC.)

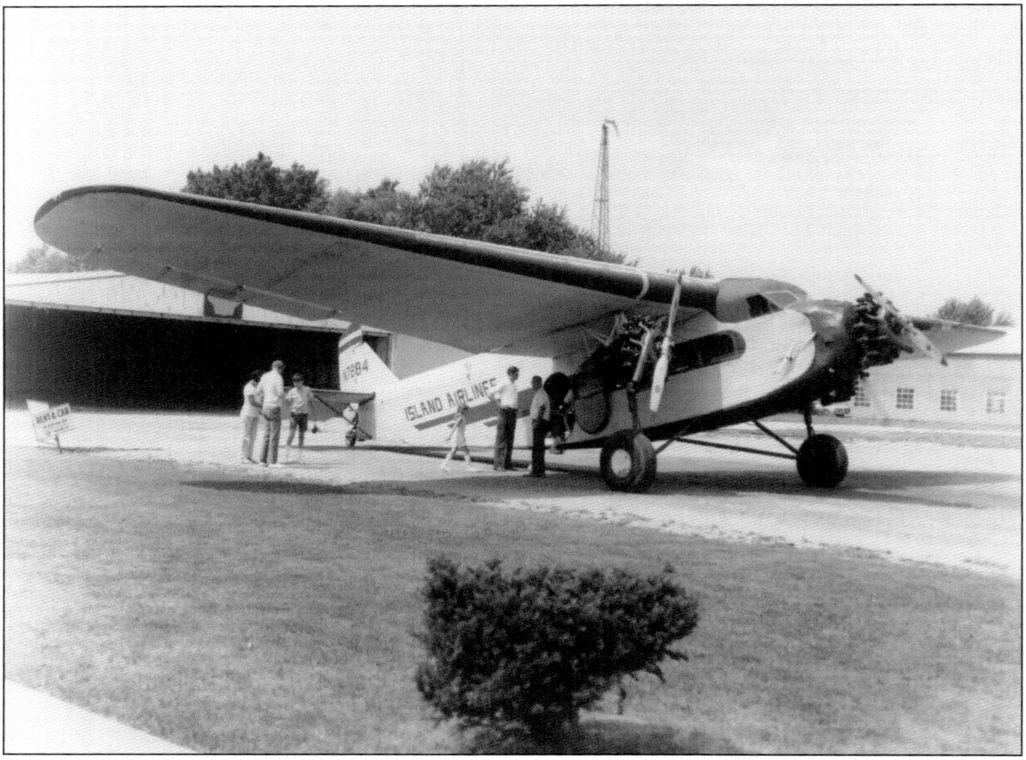

The Ford Tri-Motor, fondly known as the "Tin Goose," served both island residents and tourists, providing air transportation from Port Clinton to South Bass, Middle Bass, and North Bass Islands for 50 years. At its peak, Island Airlines utilized four Ford Tri-Motors to carry passengers and cargo to the islands. It was billed as "the world's shortest airline." Tri-Motors, like this one in the 1950s, were slow and somewhat ungraceful, but they were dependable and enjoyed a sterling safety record. The Put-in-Bay Airport has serviced South Bass Island since 1930. (Courtesy of James Lang.)

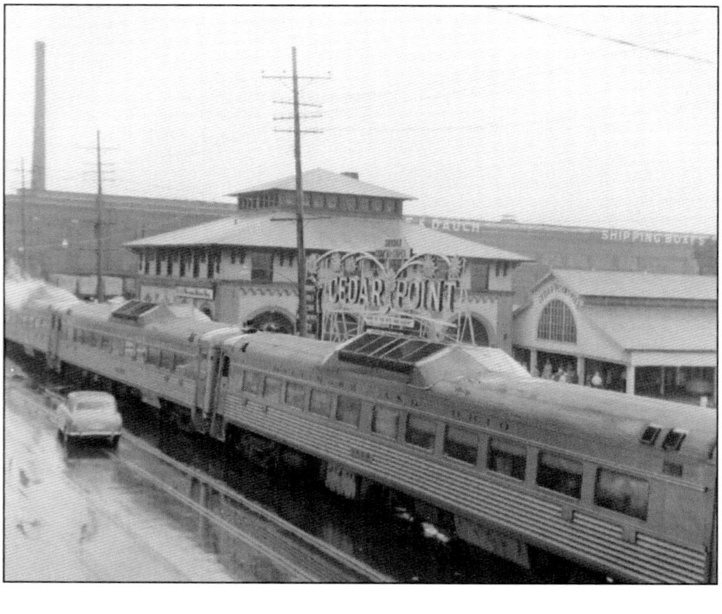

A Baltimore & Ohio train delivers passengers to the Cedar Point dock at the foot of Columbus Avenue sometime between 1955 and 1965. Train travel to Cedar Point declined significantly in the 1950s, and the tracks providing passenger service to downtown were removed in the 1970s. The Cedar Point sign at the dock, illuminated at night, was a Cedar Point icon from the 1920s until the 1980s. (Courtesy of SLARC.)

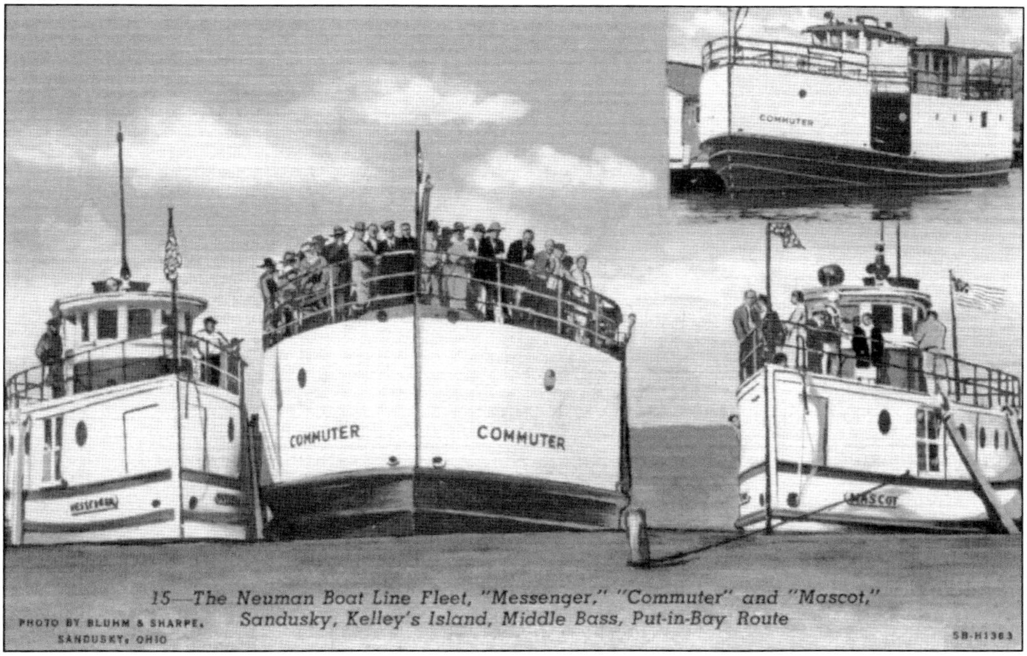

In its 1950s and 1960s heyday, the Newman Boat Line provided regular ferry service to Kelleys, South Bass, and Middle Bass Islands. This 1953 postcard shows the Newman lineup of ferryboats, which transported passengers, vehicles, and cargo. Newman Boat Line operated as a family business from 1907 until it was sold in 2001. (Courtesy of Frank Landl.)

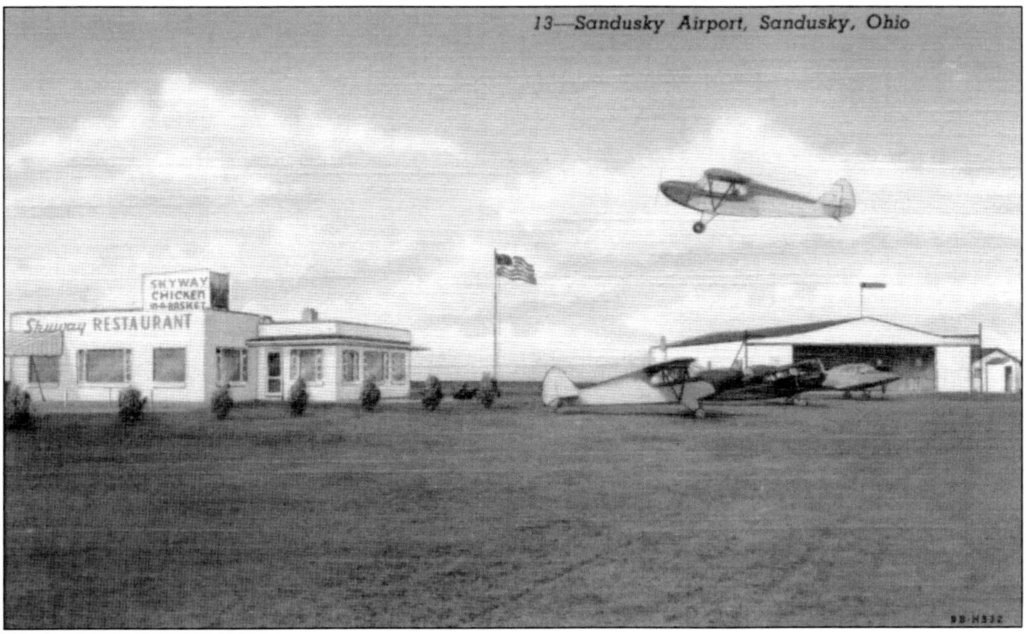

For more than 70 years, the Griffing Sandusky Airport provided service to the Lake Erie islands. Small planes, as seen in this 1950s postcard, carried mail and passengers, including children who attended school in Sandusky but lived on one of the islands. The airport closed in December 2013, and operations were moved to the Erie-Ottawa Regional Airport in Port Clinton. (Courtesy of Frank Landl.)

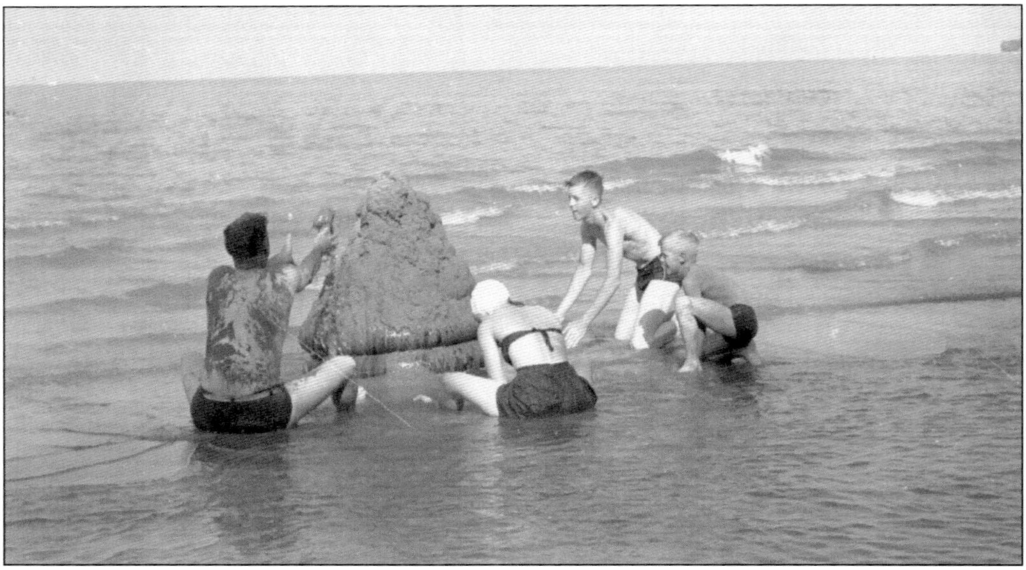

The 1950s was a banner decade for the growth and development of the state park system in Erie and Ottawa Counties, including the opening of East Harbor State Park on the Marblehead Peninsula. The park is one of the jewels of the Lake Erie shoreline. Its beautiful beach continues to be a major attraction. This c. 1955 photograph shows a family working on a sand castle. The southwestern tip of Kelleys Island, called Carpenter Point, is visible in the right background. (Courtesy of HPC.)

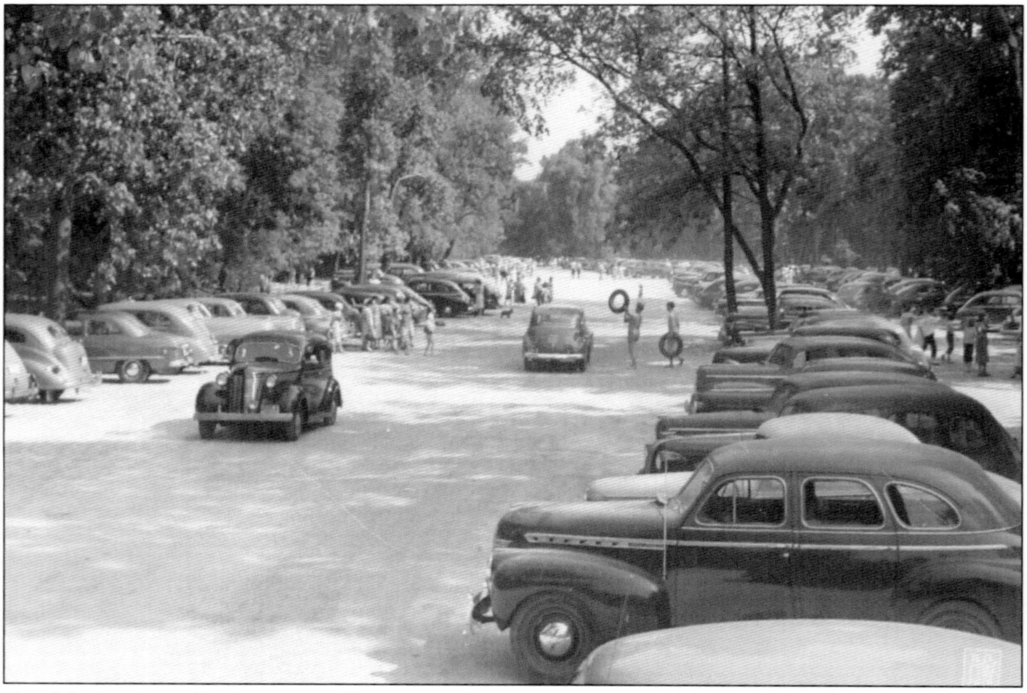

East Harbor State Park attracted huge crowds on warm summer weekends, as this 1950s lineup of vehicles attests. Automobile travel expanded quickly in the 1950s as the economy grew and roads improved. The cities that provided the region with many of its visitors, like Cleveland, Detroit, and Cleveland, were at their peak in the 1950s and 1960s. (Courtesy of HPC.)

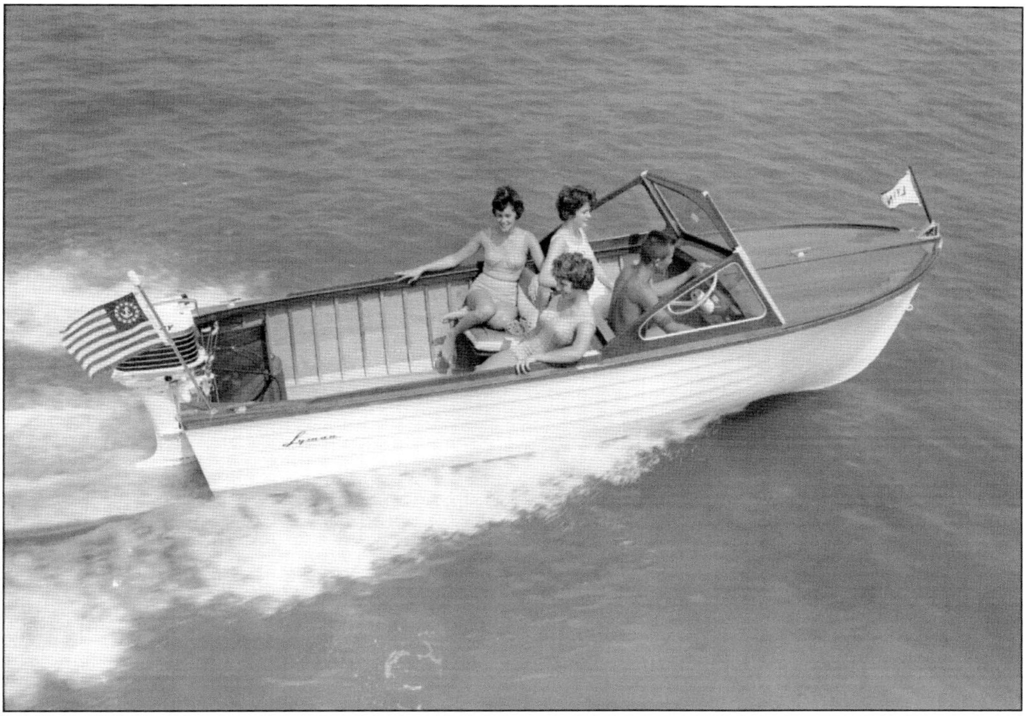

Lyman boats were some of the most popular craft on Lake Erie from the 1940s to the 1960s. Founded in 1875, the company's manufacturing process was moved to Sandusky in 1932. Lyman boats achieved their height of popularity in the 1950s and early 1960s. The clinker-built lapstrake hulls provided a stable ride on choppy waters. The growing popularity of fiberglass boats hurt Lyman sales, and the last wooden Lymans were built in the early 1970s. (Courtesy of SAMA.)

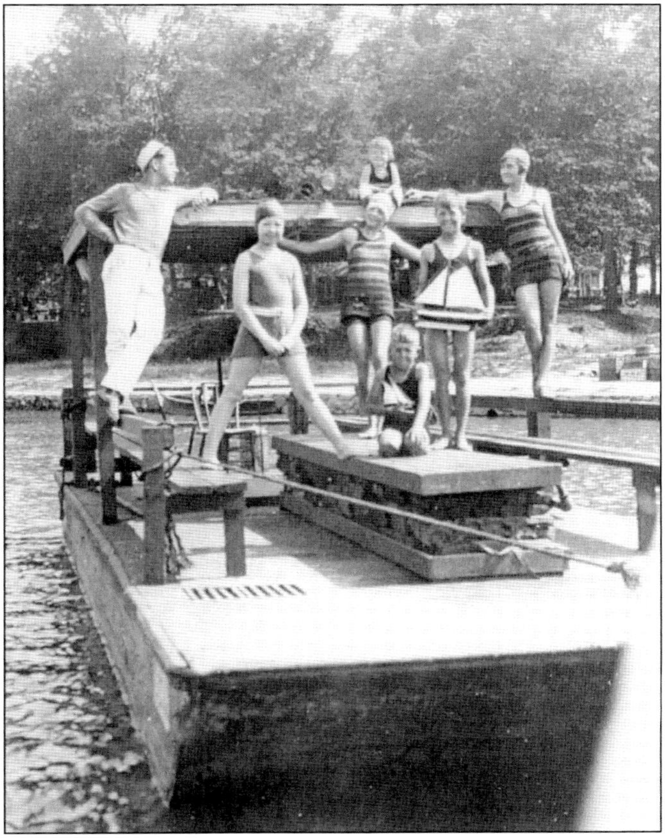

Marblehead's Channel Grove Marina has been in operation for 90 years. Seen here around 1950, owner Skip Brown (far left) poses with bathers headed to Sandy Beach, now part of East Harbor State Park. (Courtesy of Brown family archives.)

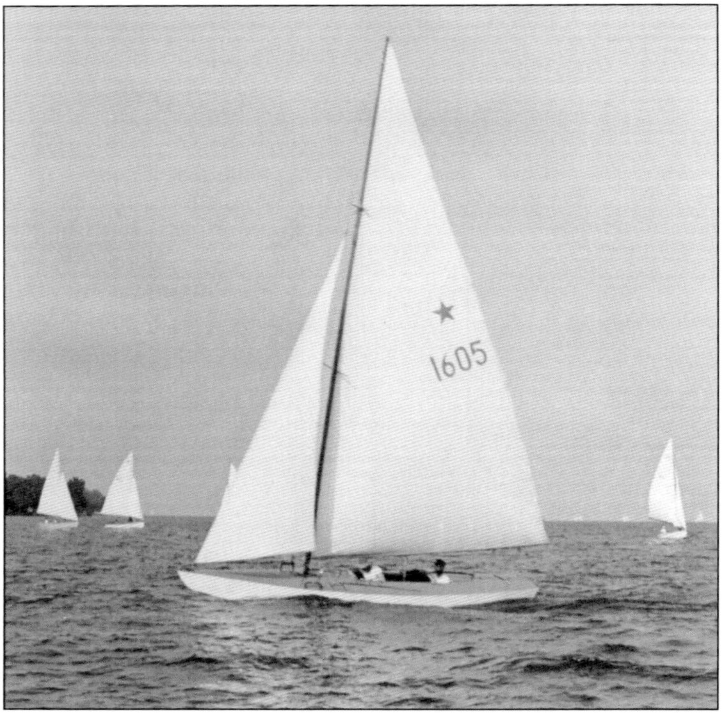

Lake Erie has always been a challenge for sailors. Here, two crew members participate in a Star class sailboat race around 1960 off the Vermilion River. Vermilion housed a large fleet of Star sailboats in the 1960s. At the time, the Star was one of the classes used for Olympic competition. (Courtesy of Ohio History Connection.)

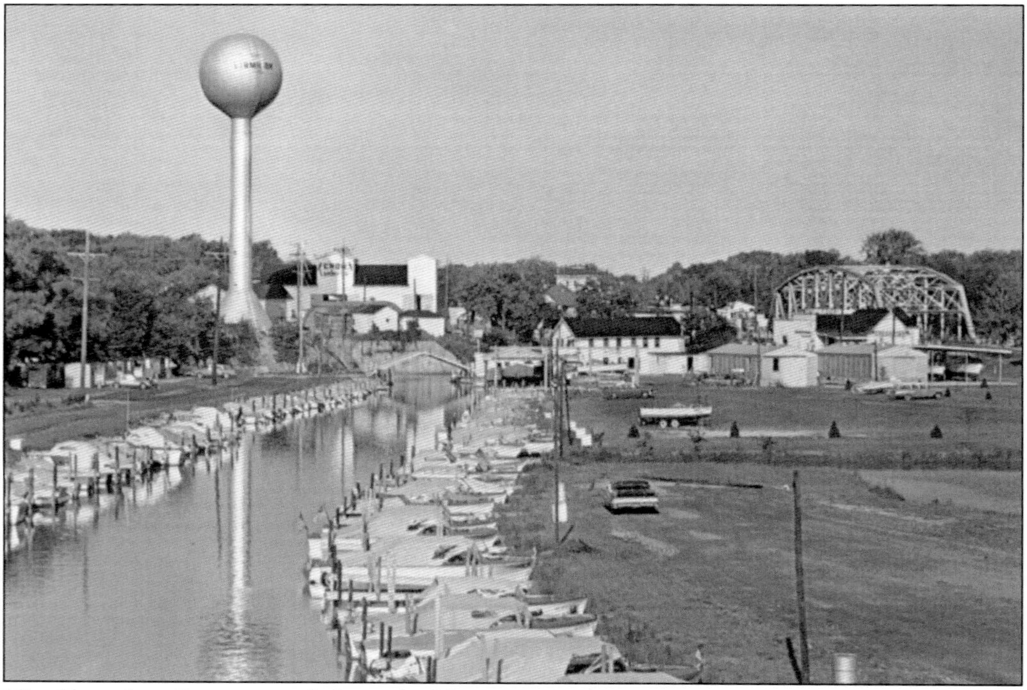

The Vermilion River provided an ideal location for marina development and boat dockage, including winter storage. In this postcard photograph from about 1960, the famous Vermilion Water Tower, a local landmark for boaters, and the US 6 highway bridge frame the background. Vermilion attracted many boaters from the Cleveland area and marked the unofficial start of the shores and islands vacation region. (Courtesy of Frank Landl.)

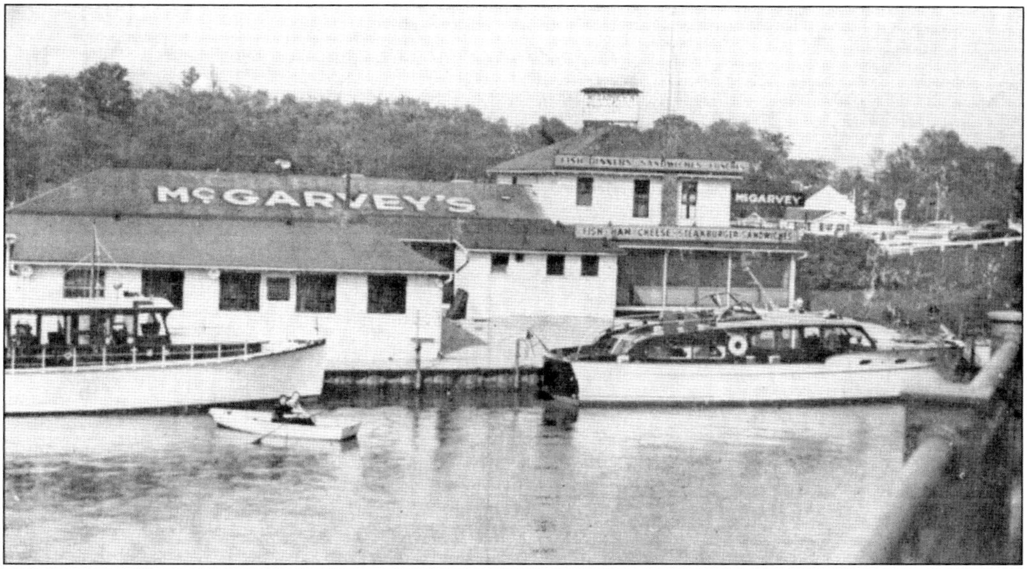

McGarvey's Restaurant, just north of the US 6 bridge over the Vermilion River, was one of the most visited restaurants on the Lake Erie shore for decades. The Solomon family operated the restaurant for 45 years, starting in 1944. A succession of restaurants has operated at the location since the late 1980s. This 1950s photograph shows the accessibility of the restaurant by boat or automobile. Spring floods were a constant concern for the owners of the restaurant. (Courtesy of Vermilion Archival Society.)

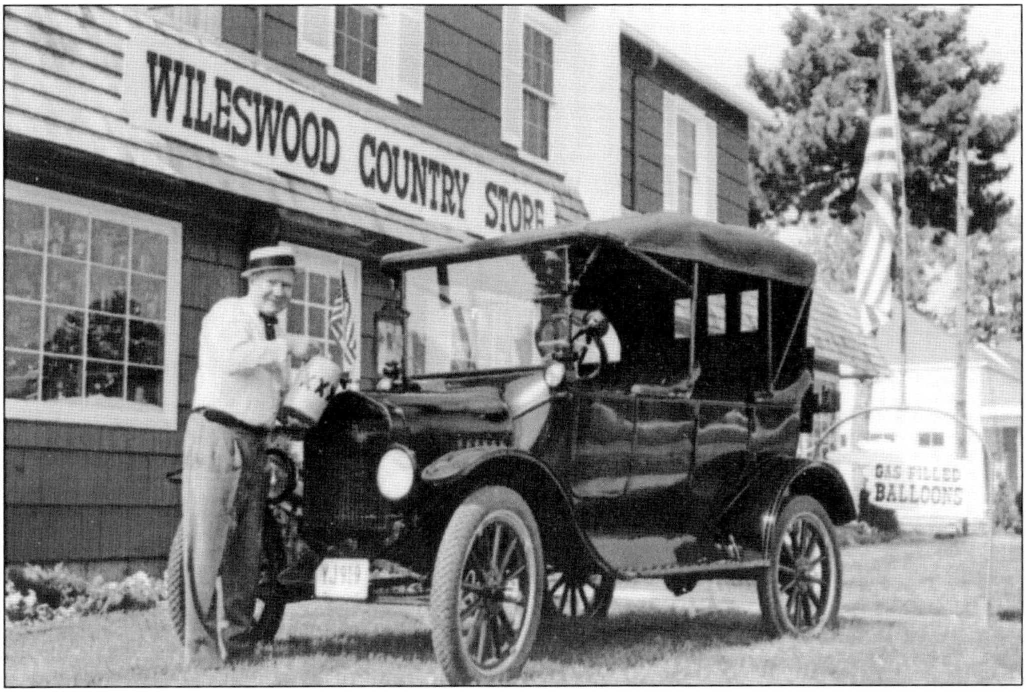

Addison Wiles, founder of Wileswood Country Store in Huron, poses in front of the store in 1963. Wiles modeled his store after a New England general store. Its most popular offerings were popcorn and penny candy. Located on US 6 just over the Huron River Bridge, the store closed in the 1990s. (Courtesy of Sheila Ehrhardt.)

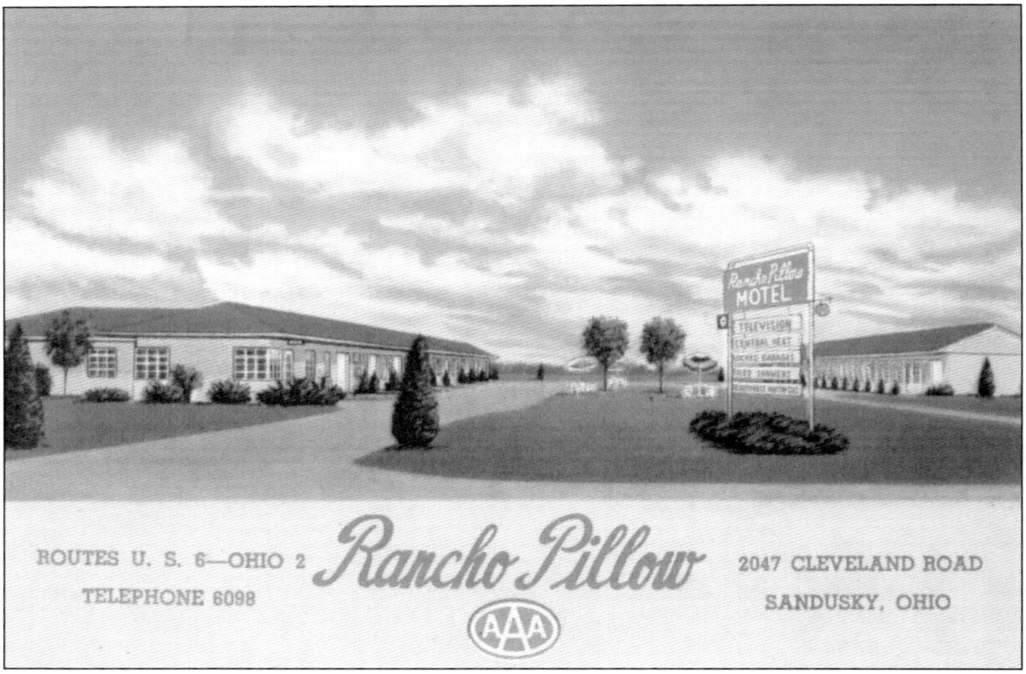

The 1950s was a decade of growth for most tourist businesses in the area. A number of independent motels and hotels were built along major highways, primarily targeting the automobile traveler. The Rancho Pillow catered to Cedar Point visitors. (Courtesy of Frank Landl.)

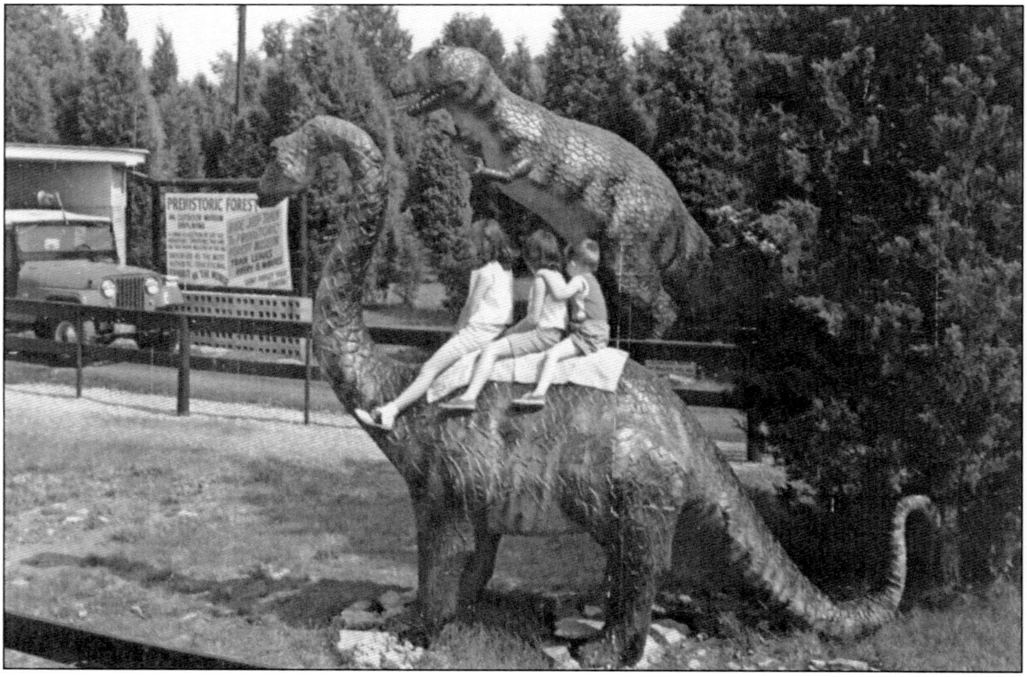

One of the classic roadside attractions on the Marblehead Peninsula was Prehistoric Forest and Mystery Hill, in operation from 1954 to 2011. Children had the opportunity to meet dinosaurs and experience optical illusions. This photograph dates to the 1950s or 1960s. (Courtesy of Frank Landl.)

The oldest lighthouse in continuous operation on the Great Lakes, the Marblehead Lighthouse was built in 1822. Its first tender was Benjamin Wolcottt, a veteran of the Revolutionary War. One of Lake Erie's most photographed sites, it marks the northeast corner of the Marblehead Peninsula and guards the entrance to Sandusky Bay. The lighthouse, seen here in the 1950s, is now operated as a state park. (Courtesy of Jack Coffman.)

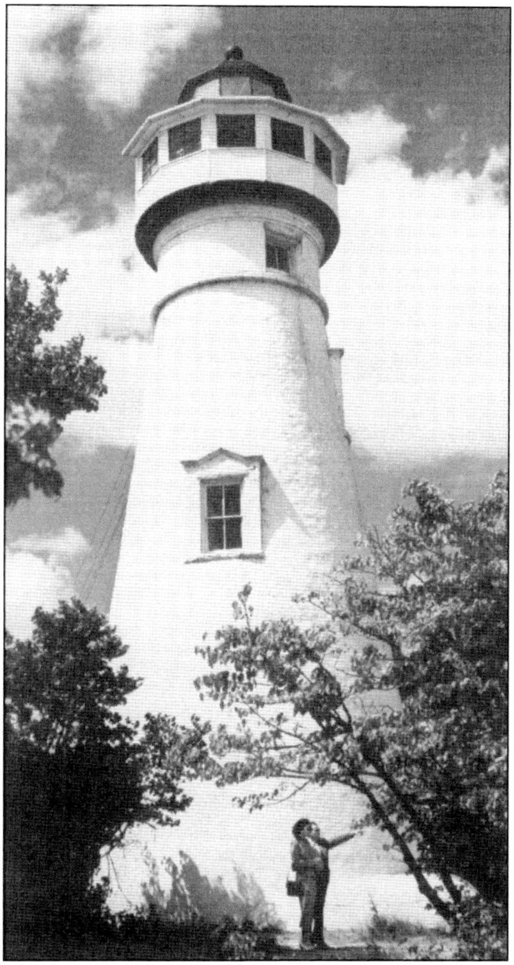

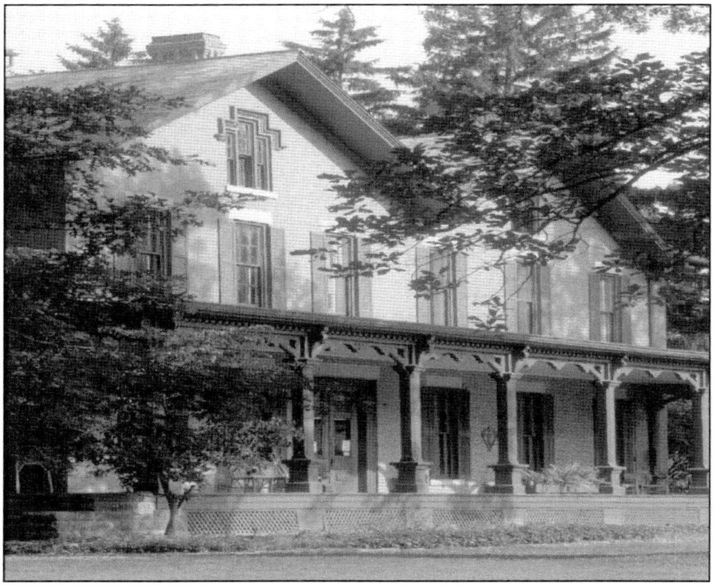

The Rutherford B. Hayes Presidential Center in Fremont includes a museum, library (the first presidential library, it opened in 1916 and expanded in 1968), and the 31-room Victorian mansion of Hayes and his wife, Lucy. The 19th president of the United States, Hayes was elected in 1876 and served one term. Spiegel Grove, the name Hayes gave his estate, is visited by thousands each year and hosts many special events. (Courtesy of HPC.)

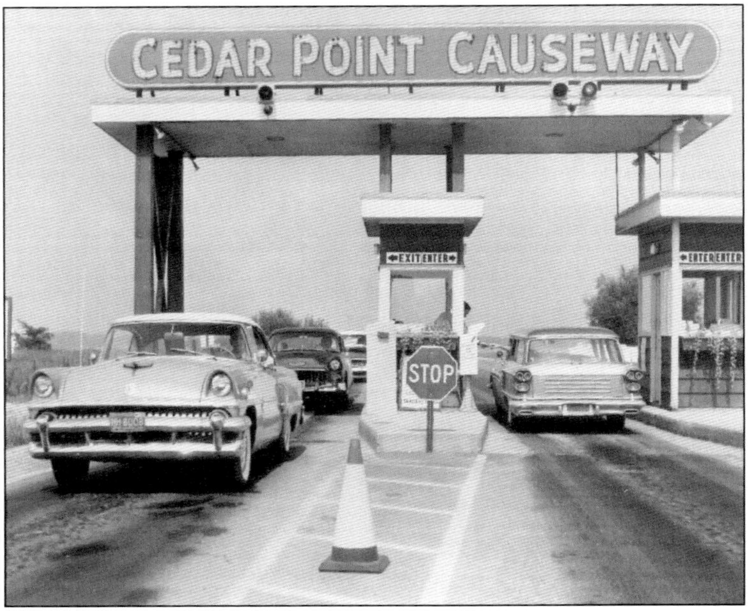

The Cedar Point Causeway, which opened in 1957, dramatically increased the convenience for guests who visited the park by automobile. For the first 40 years of its existence, Cedar Point visitors arrived by boat. Now, just about all arrived by automobile. (Courtesy of CPA.)

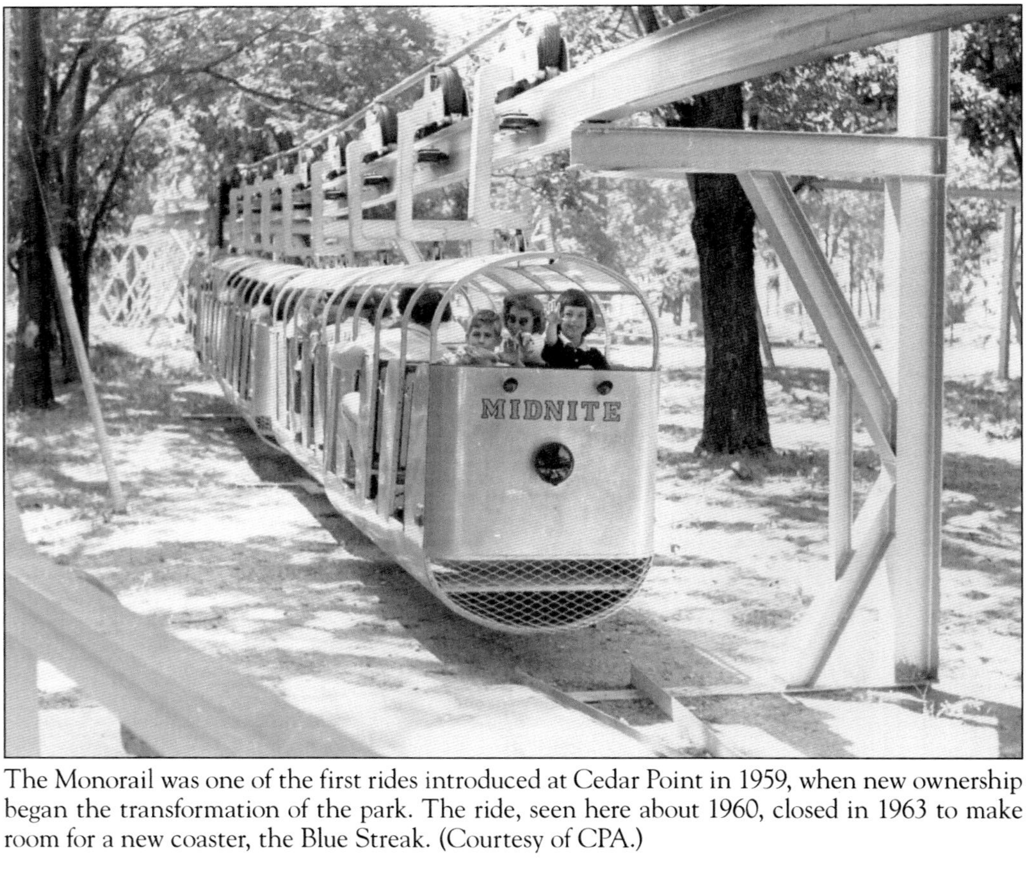

The Monorail was one of the first rides introduced at Cedar Point in 1959, when new ownership began the transformation of the park. The ride, seen here about 1960, closed in 1963 to make room for a new coaster, the Blue Streak. (Courtesy of CPA.)

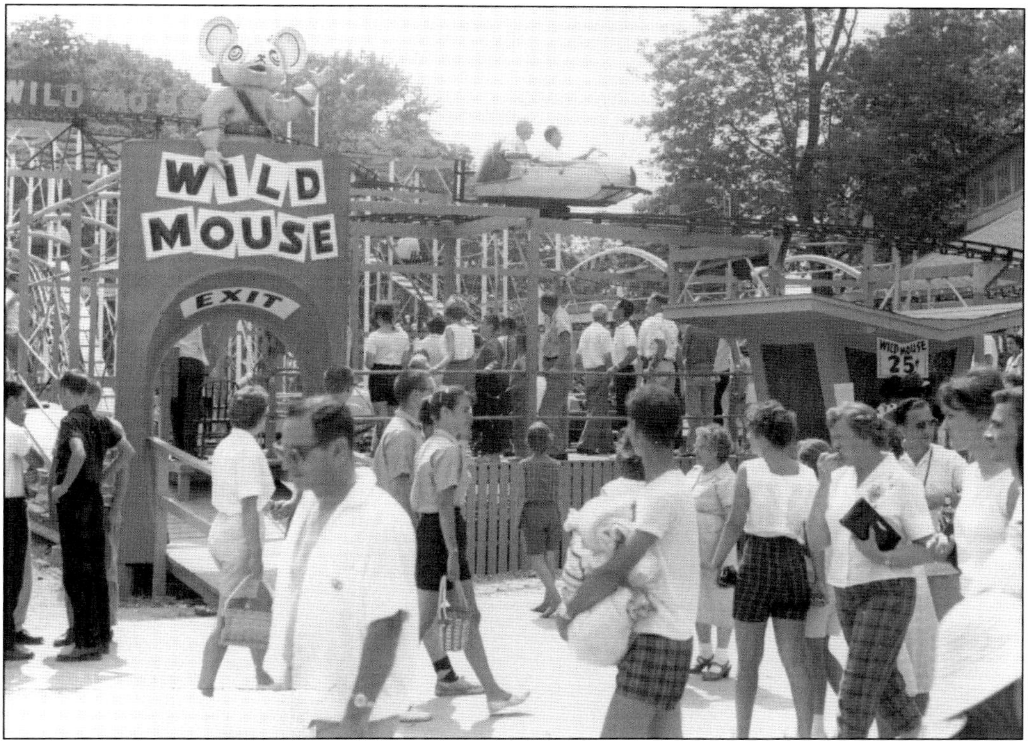

The Wild Mouse, which opened in 1959, was Cedar Point's first steel roller coaster. It operated through the 1962 season. Built by B.A. Schiff & Associates, it packed lots of thrills into a small space, giving riders lots of air time as they went up and down the hills. Rides similar to the Wild Mouse continue to operate in amusement parks around the world. (Courtesy of CPA.)

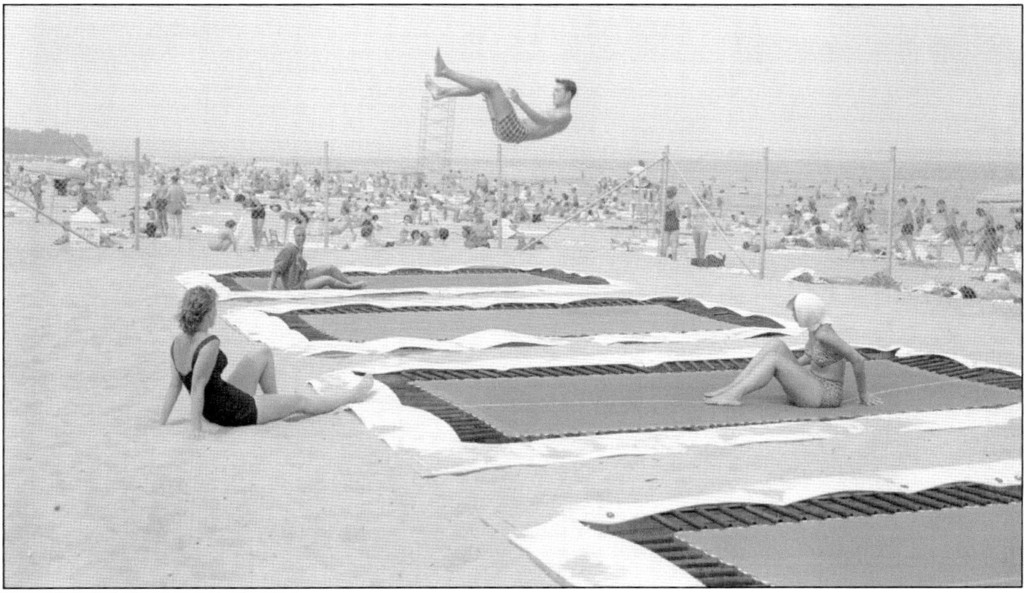

A young acrobat bounces high off a trampoline on the Cedar Point Beach around 1960. The beach was still popular with guests in the 1960s, but it was beginning to lose some of its appeal to the midway as new rides and attractions were added every year. (Courtesy of CPA.)

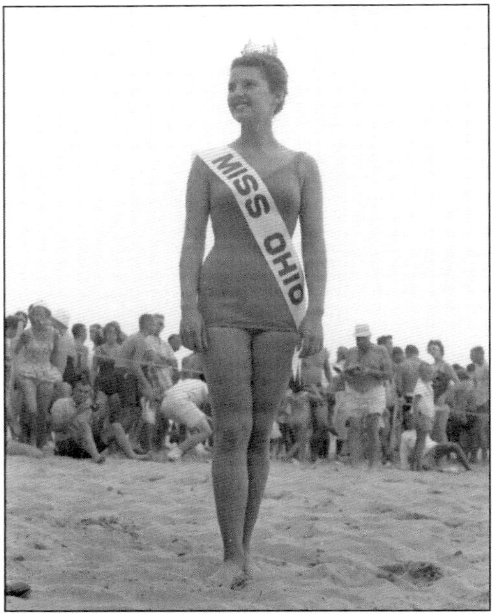

Celebrated Sanduskian Jackie Mayer was crowned Miss Ohio in 1962 and Miss America in 1963. Here, she is being photographed on the Cedar Point Beach in the summer of 1962. Catching a look at a beauty queen never loses its appeal. (Courtesy of Jackie Mayer.)

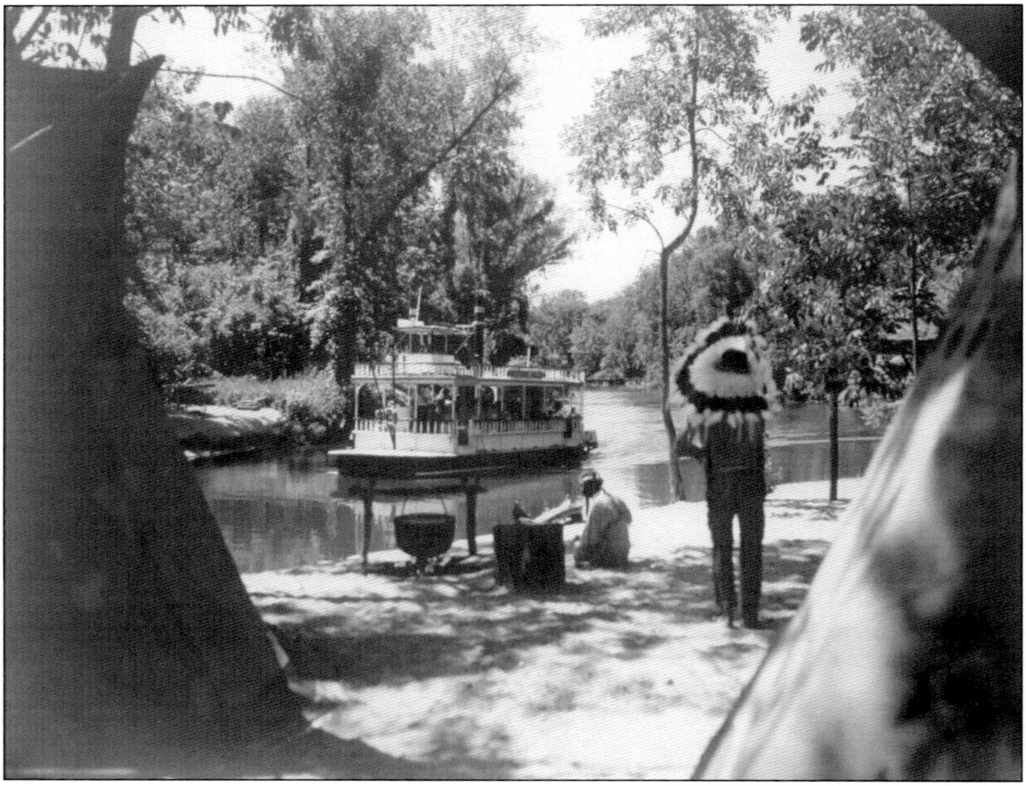

The canals and lagoons dug by George Boeckling in 1905 were put to a new use in 1964, when the Western Cruise opened at Cedar Point. Guests were taken on a boat ride through the early history of Ohio, complete with animated Indians, snarling cougars, and frontier settlers. The ride operated at the park through the 2011 season, when it was retired to make room for the Dinosaurs Alive attraction. (Courtesy of HPC.)

The Blue Streak, named for the Sandusky High School athletic teams, opened in 1964. It was the first new wooden roller coaster at the park since the Cyclone in 1929. A classic out-and-back coaster with a fast and hilly ride, the 78-foot-high Blue Streak is a longtime guest favorite. It celebrated its 50th anniversary in 2014. (Courtesy of HPC.)

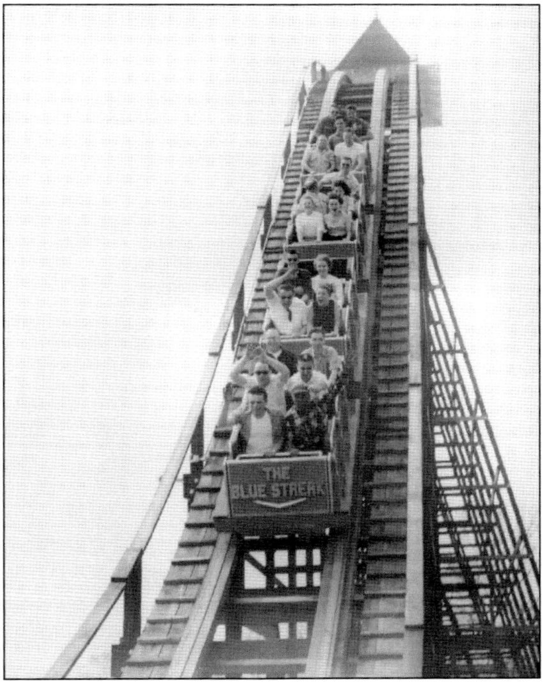

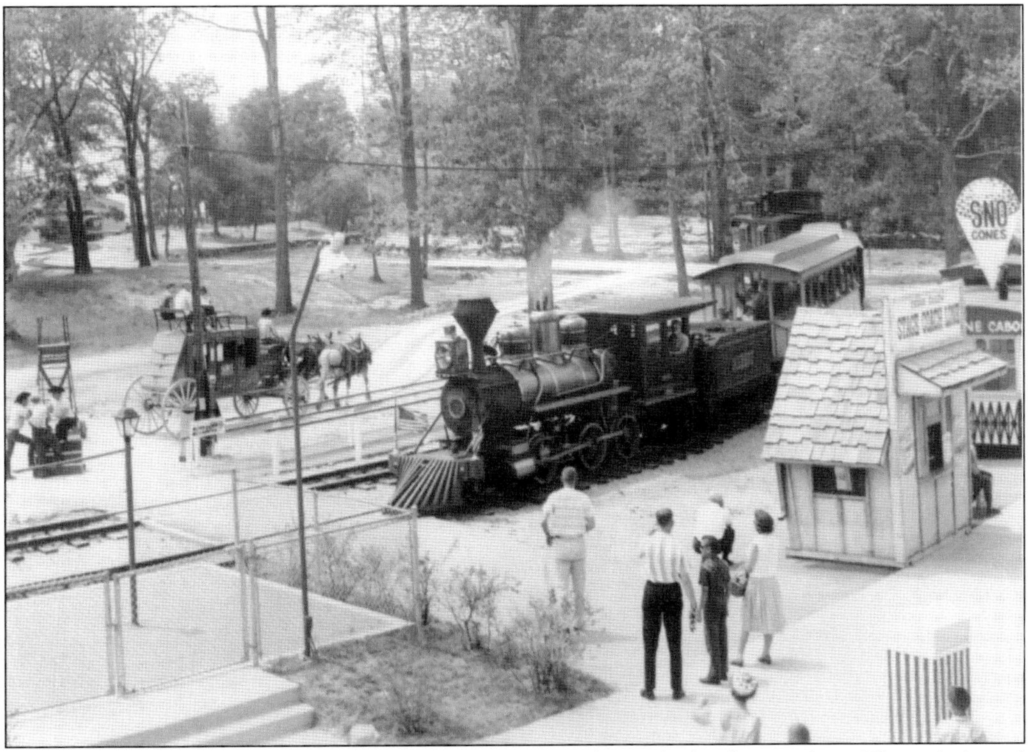

The Stagecoach ride at Cedar Point included simulated attacks by desperadoes. The ride lasted only two seasons, closing in 1968. The Cedar Point & Lake Erie Railroad, which opened in 1963, is the most popular ride in the history of Cedar Point. It features authentic steam-powered, coal-burning locomotives. It celebrated its 50th anniversary in 2013. (Courtesy of CPA.)

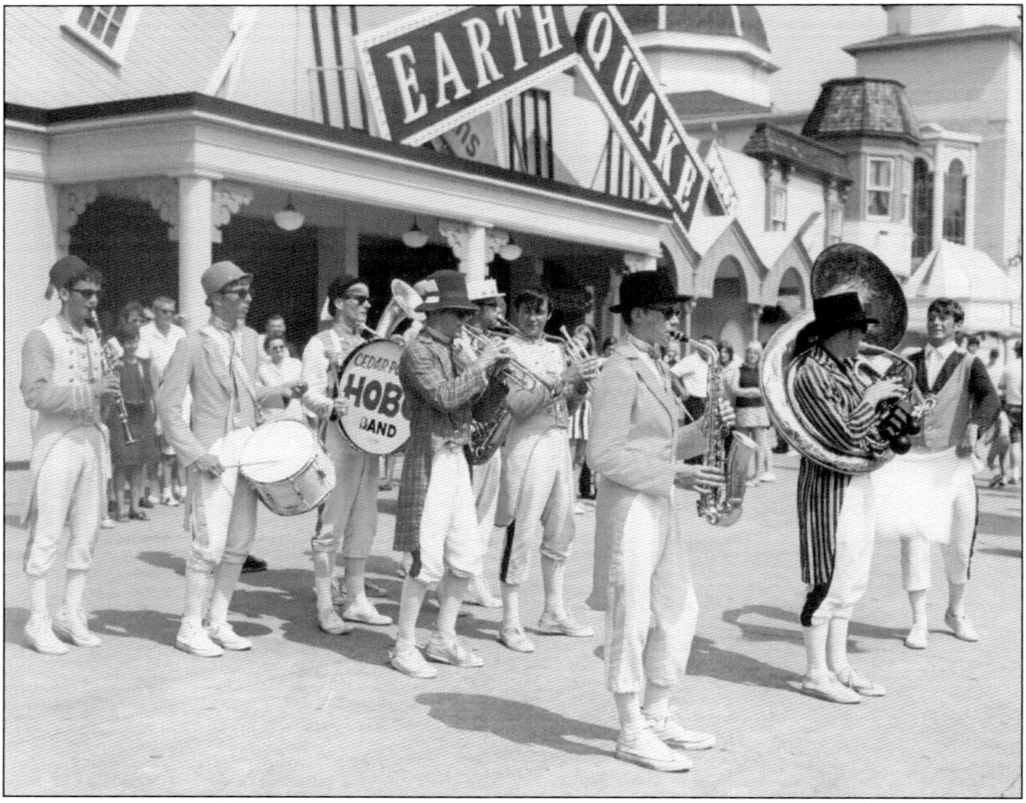

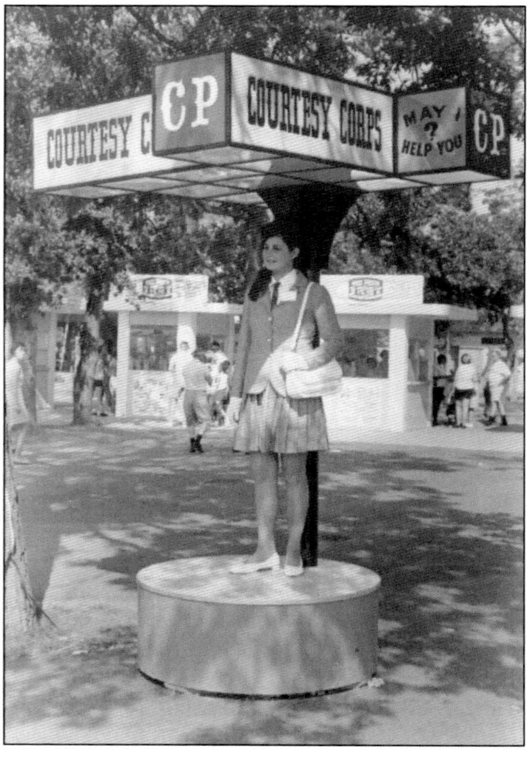

Cedar Point's famous Hobo Band plays on the main midway around 1968. Cedar Point recruited college musicians to staff its live entertainment department. The Hobo Band roamed the midways, playing impromptu sessions for guests, including a few bits of comedy. (Courtesy of CPA.)

One element of the park's new guest-centered operating philosophy (freely borrowed from Disneyland) was the creation of the Courtesy Corps, a group of college girls recruited for their guest service skills. Their role was to answer questions and provide information or assistance to guests. A member of the Courtesy Corps is seen here around 1968. (Courtesy of CPA.)

The late Lawrence "Jungle Larry" Tetzlaff was a fixture at Cedar Point for 19 seasons, starting in 1965 and continuing until his death in 1984. Along with his wife, Nancy, and later with their sons David and Tim, the family's Jungle Larry's African Safari offered wild animal exhibits and show acts involving lions, tigers, and chimpanzees. Following the 1994 season, the attraction moved permanently to Naples, Florida, where it had maintained a winter operation for many years. Here, Jungle Larry plays with a tiglon (half tiger, half lion) around 1969. (Courtesy of CPA.)

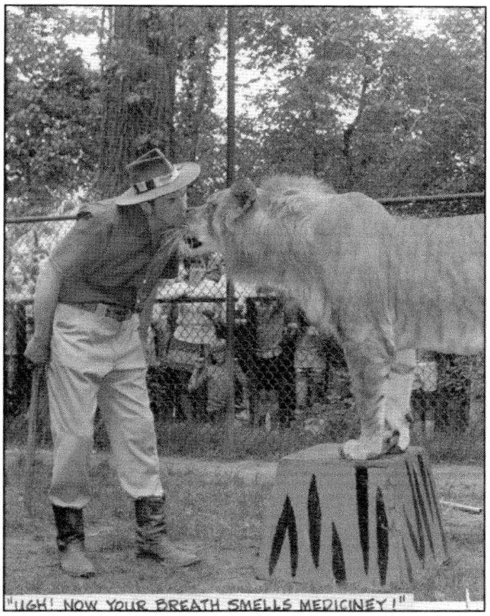

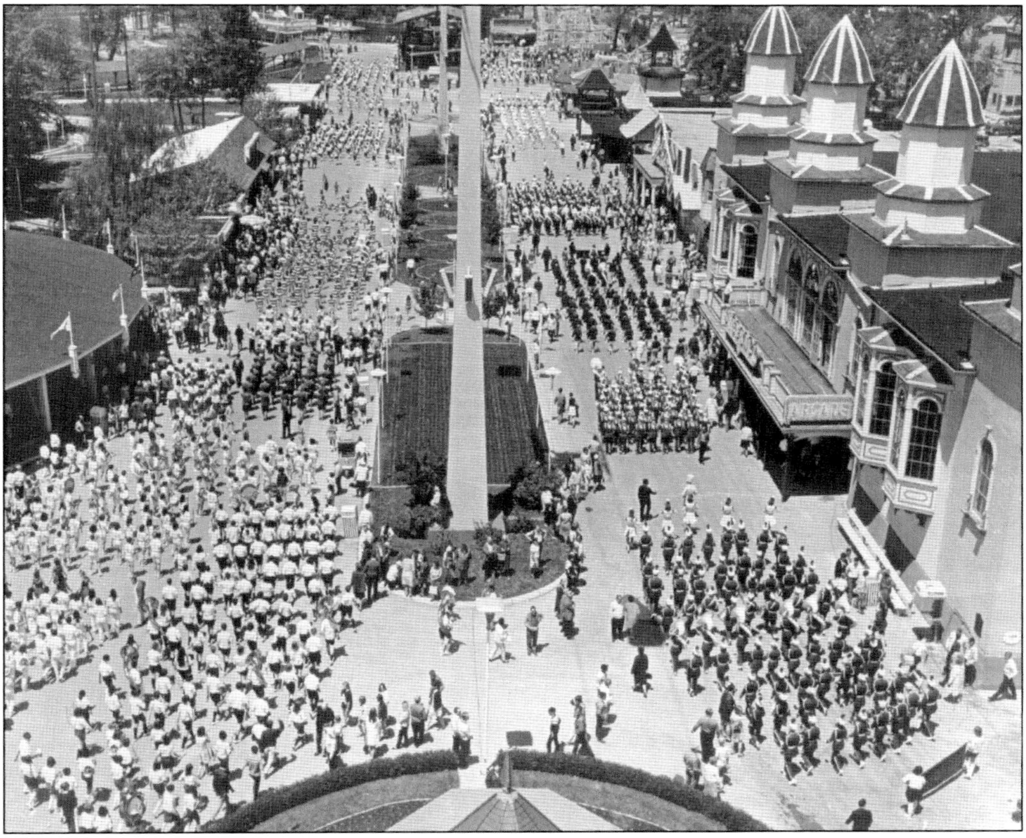

Dozens of high school bands from all over Ohio play on opening day at Cedar Point around 1969. The bands were an opening day tradition in the 1960s and 1970s. This photograph showcases many of the changes at Cedar Point made by the Roose and Legros ownership: a new concrete midway, Sky Ride, Western Cruise, large floral displays, Wild Mouse coaster, and the refurbished Coliseum. (Courtesy of HPC.)

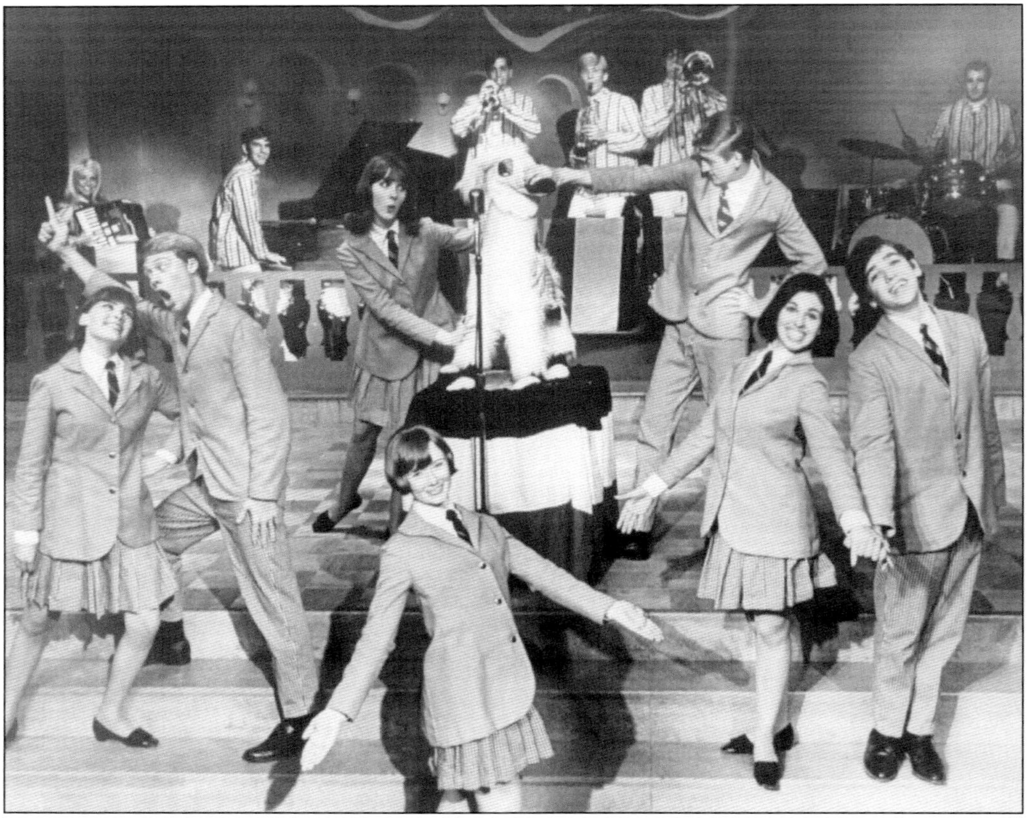

Cedar Point's live entertainment offering has always been varied. In the 1960s, the park offered Broadway-style revues, strolling acts, Dixieland bands, and country acts, all performed by college students studying music or theater. This photograph dates to around 1968. (Courtesy of CPA.)

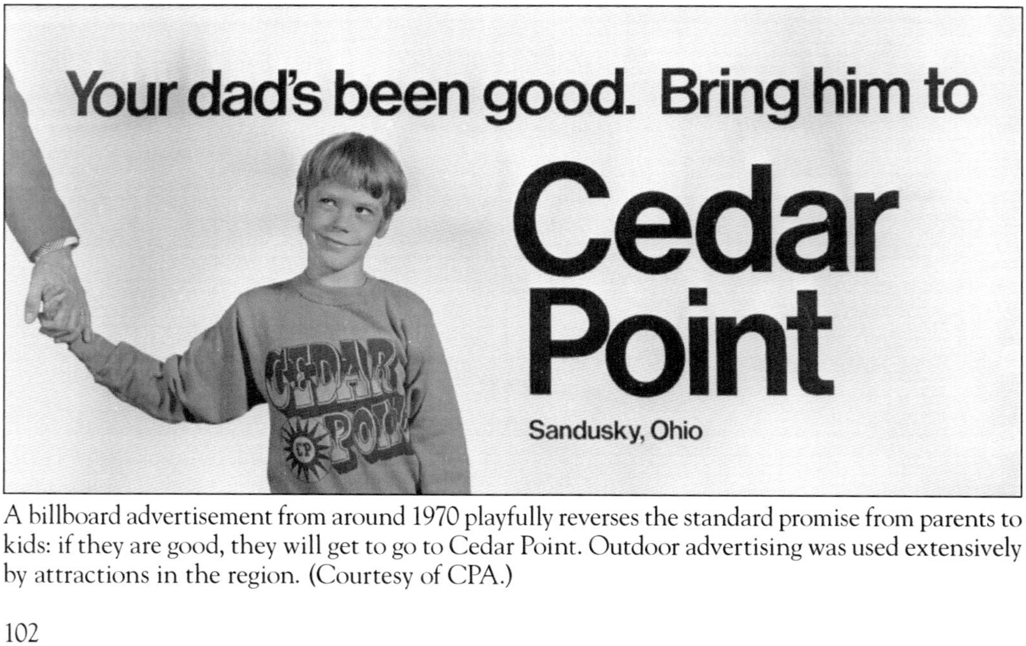

A billboard advertisement from around 1970 playfully reverses the standard promise from parents to kids: if they are good, they will get to go to Cedar Point. Outdoor advertising was used extensively by attractions in the region. (Courtesy of CPA.)

Six

Lake Erie Renaissance 1970–1990

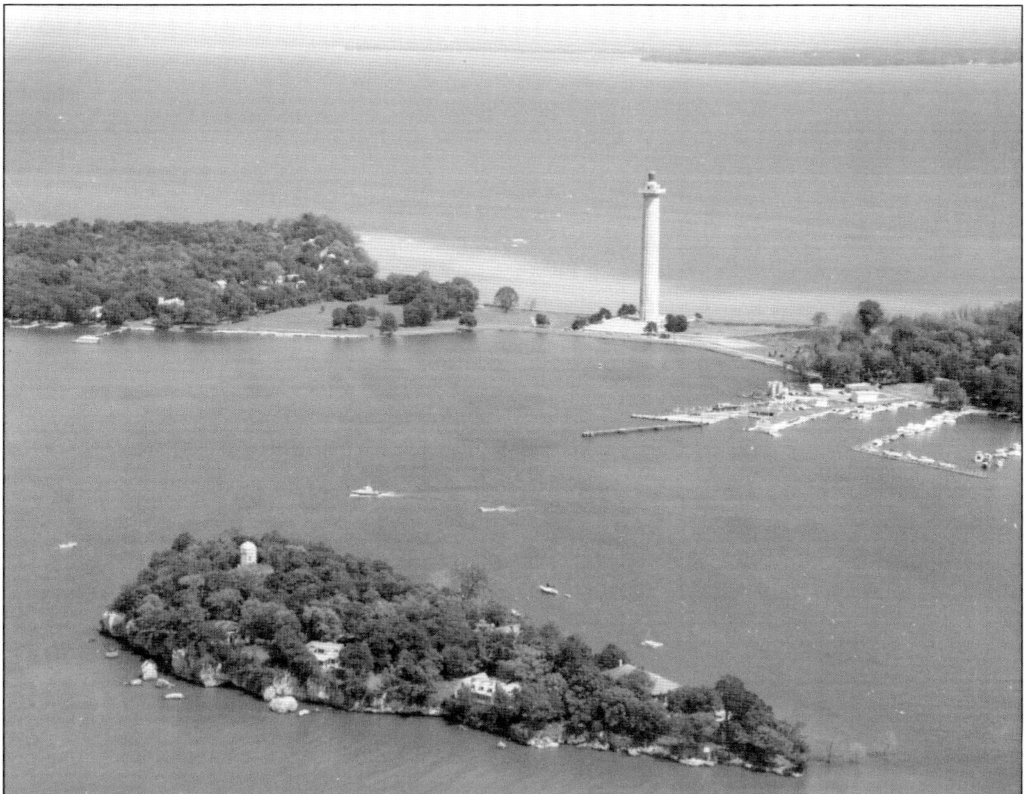

Gibraltar Island (foreground), once the summer home of Civil War financier Jay Cooke, is now home to The Ohio State University's Stone Laboratory, the oldest freshwater biological field station and research laboratory in the United States. The 1970s was a transition decade for Lake Erie. The pollution concerns of the 1960s were mitigated with new waste treatment facilities and the reduction of phosphorus in the lake. By the end of the decade, the lake had made a dramatic recovery. (Courtesy of Ohio History Connection.)

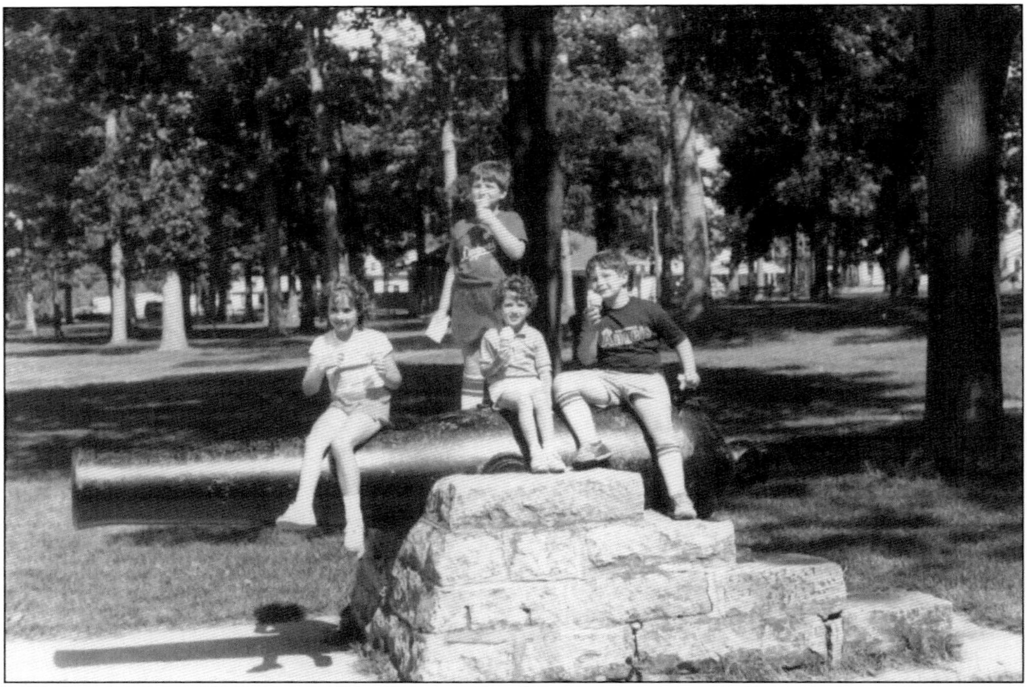

Children enjoying ice cream cones pose on an 1812-era cannon located in De Rivera Park in downtown Put-in-Bay around 1986. The park is named for the businessman and developer who owned much of South Bass Island in the middle of the 19th century. De Rivera deeded the park property to the Village of Put-in-Bay with the stipulation that it was to remain "ever free for the people." (Courtesy of Melissa Martilotta.)

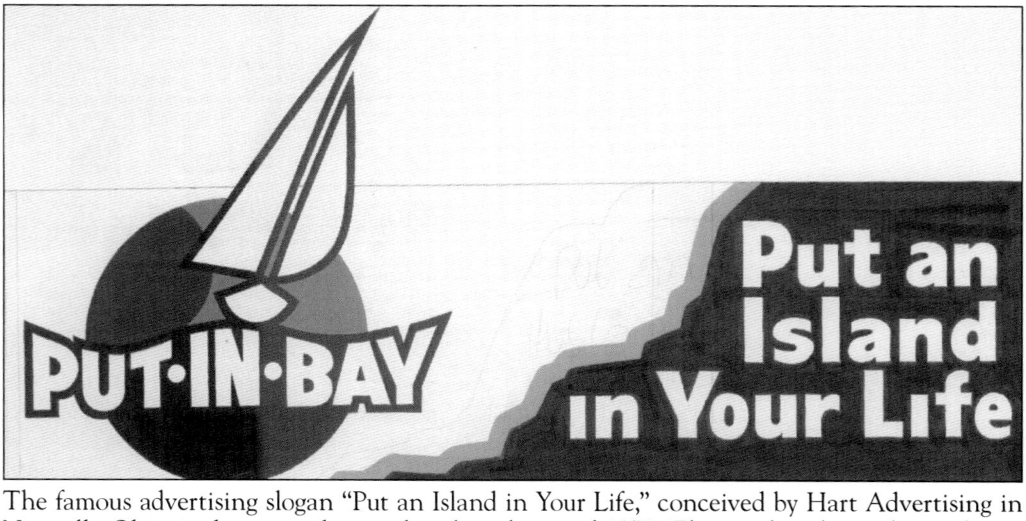

The famous advertising slogan "Put an Island in Your Life," conceived by Hart Advertising in Norwalk, Ohio, is shown in this outdoor board around 1972. The graphics have changed over the years, but the message has not. (Courtesy of Hart Advertising.)

A freewheeling, Key West–like atmosphere at Put-in-Bay, especially in the downtown bars and restaurants, developed throughout the 1970s and 1980s. One of the main reasons people came to Put-in-Bay was to see and hear Pat Dailey, the performer who created a special magic with his songs about island life. Dailey began performing at the Beer Barrel Saloon in 1978 and has worked on the island ever since, splitting his time between Key West and Put-in-Bay. (Courtesy of Dale Sadowsky.)

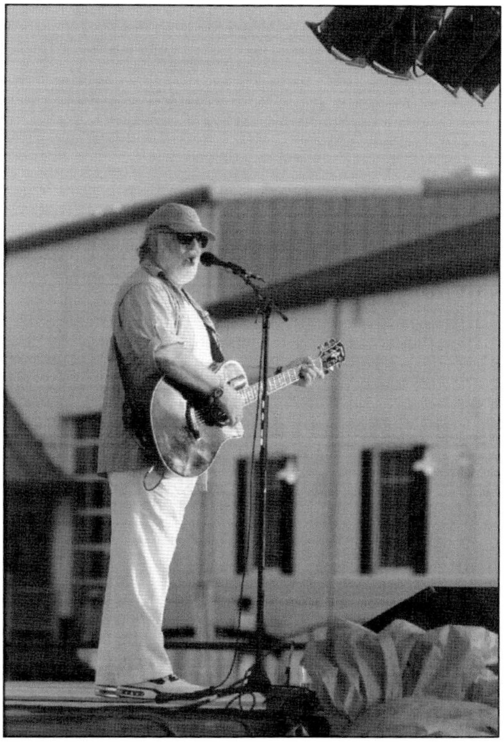

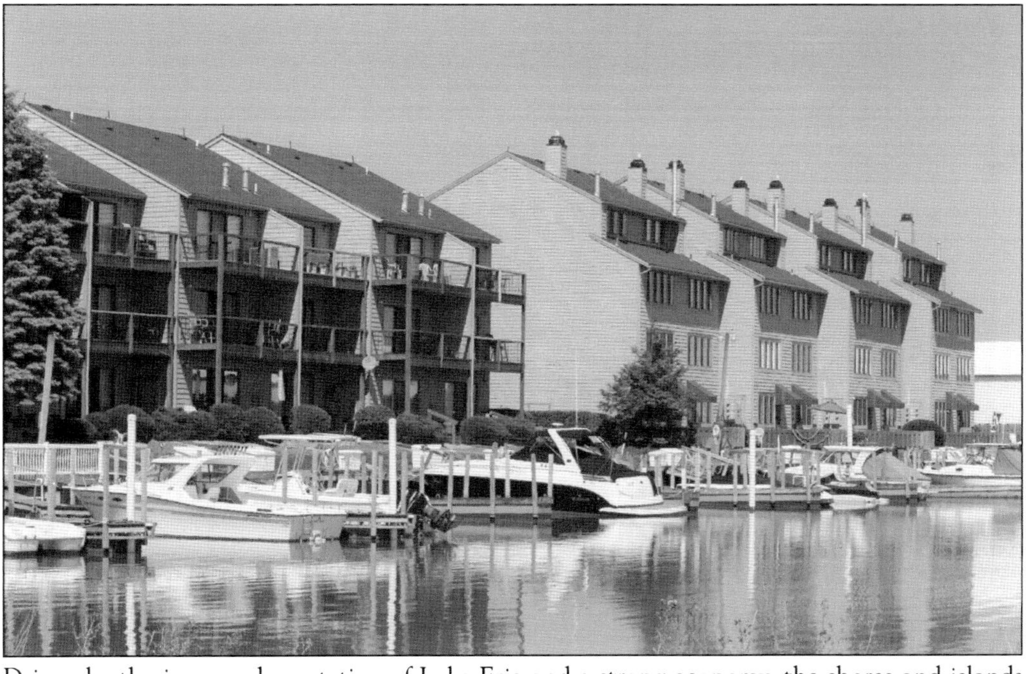

Driven by the improved reputation of Lake Erie and a strong economy, the shores and islands area benefited from the growth of second-home development in the 1980s. Condominium communities, some very upscale, were built in Sandusky, Marblehead, Catawba, and the Lake Erie islands. Pictured here is Harborside in Port Clinton. Many condo developments included boat docks. (Courtesy of LESI.)

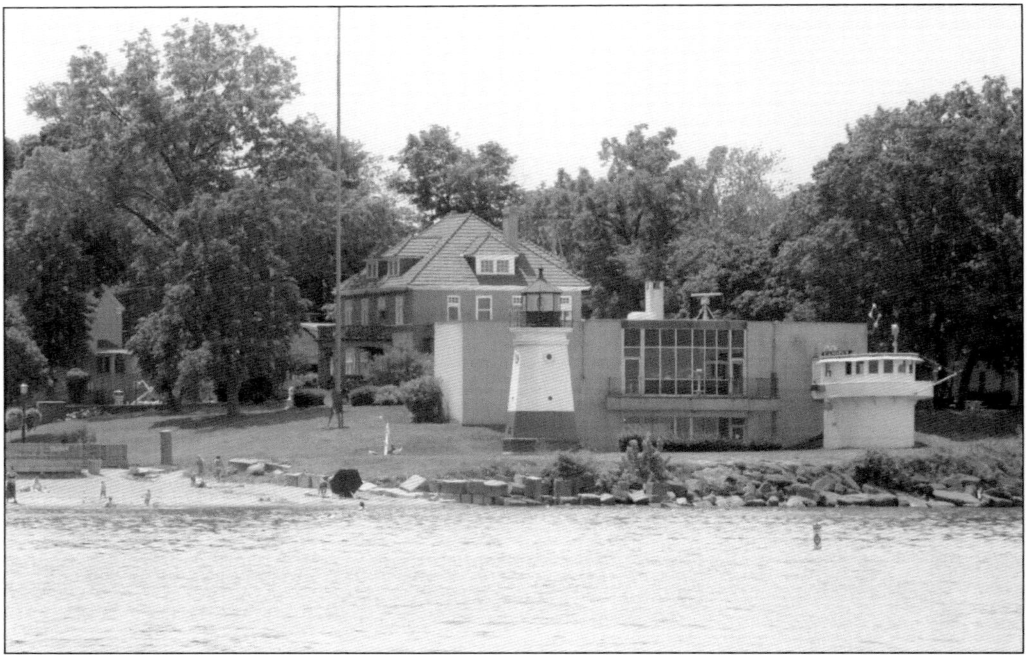

The stately 1909 brick home of Commodore Fred Wakefield was converted to the Great Lakes Historical Society Museum in 1956. One of the most popular attractions in Vermilion, the museum remained open until 2014, when it was moved to Toledo, Ohio. Today, the building continues to house the Great Lakes Society Archives. (Courtesy of Vermilion Archival Society.)

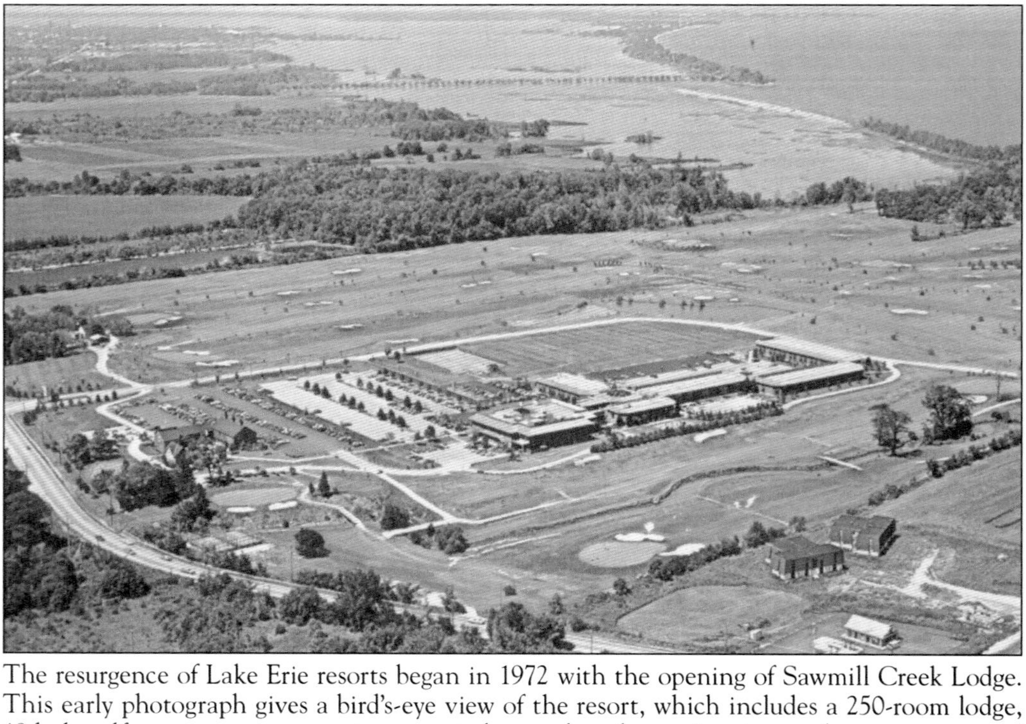

The resurgence of Lake Erie resorts began in 1972 with the opening of Sawmill Creek Lodge. This early photograph gives a bird's-eye view of the resort, which includes a 250-room lodge, 18-hole golf course, marina, tennis courts, indoor and outdoor swimming pools, restaurants, and a condominium development. (Courtesy of Frank Landl.)

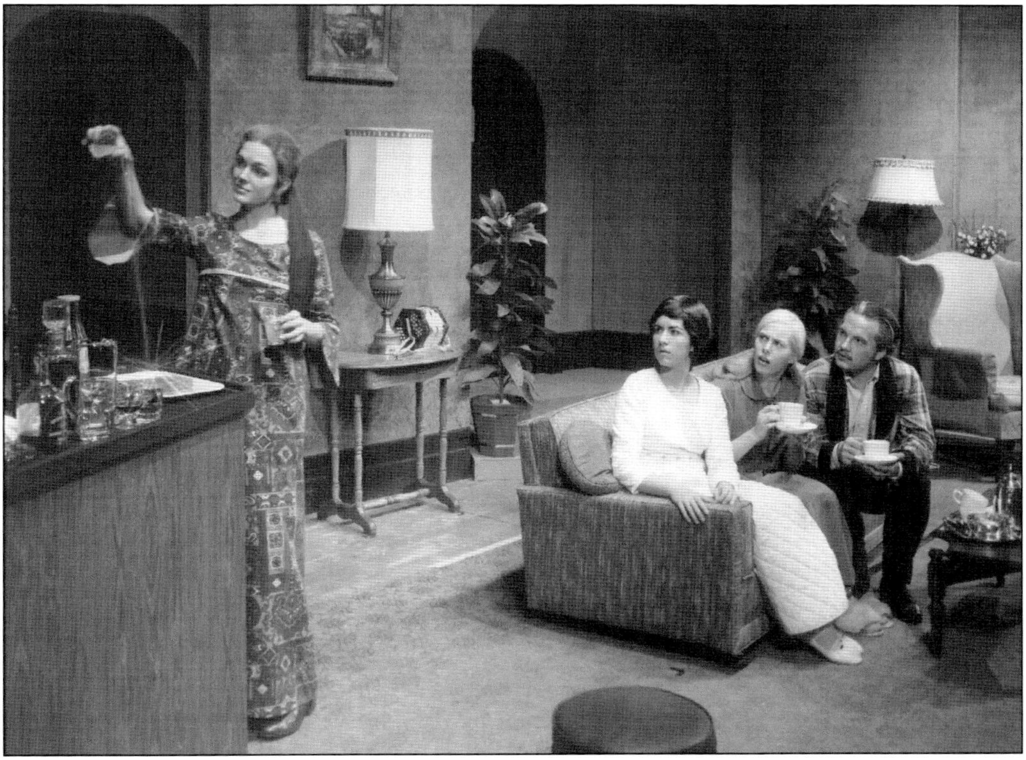

Since 1949, the Huron Playhouse has provided both summer visitors and residents with quality stage productions, including comedies, dramas, and musicals. In 1970, the cast of *A Delicate Balance* entertains theatergoers. Shown are, from left to right, Marje Rody, Helen Finch, Deborah Holt, and Steve Carr. (Courtesy of Marje Rody.)

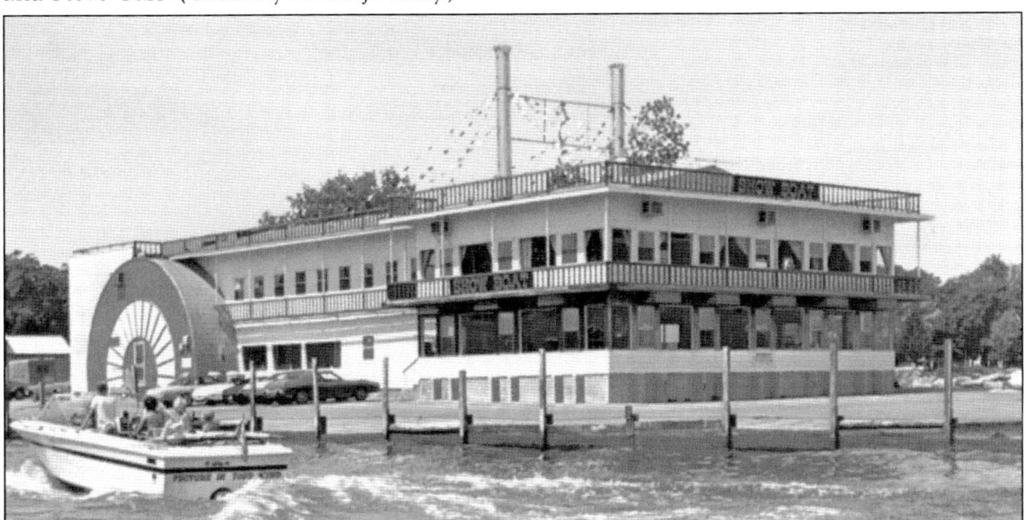

One of the most distinctive buildings in Huron in the 1970s and the 1980s was the Showboat Restaurant. Built on the Huron Pier to resemble a side-wheel steamboat, the restaurant and lounge, seen here around 1980, offered spectacular views of the Huron River and Lake Erie. When the restaurant opened in 1971, the seafood buffet was priced at $5.95. The restaurant closed in 1994 and has since been torn down. (Courtesy of Sheila Ehrhardt.)

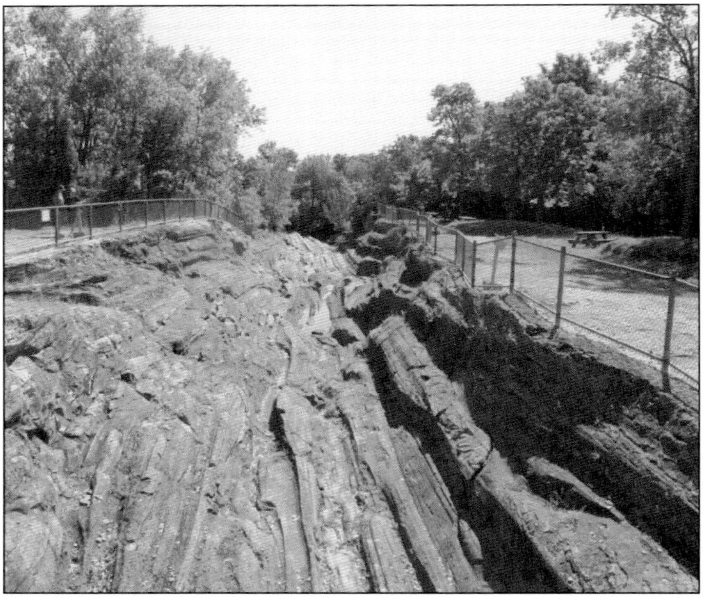

The Glacial Grooves on Kelleys Island are the largest easily accessible glacial grooves in the world. The Devonian-age limestone groove is 400 feet long, 35 feet wide, and 15 feet deep. It was created 30,000 years ago by the scraping and scouring effect of the Wisconsin glacier. The Glacial Grooves, discovered during quarrying operations, is a National Natural Landmark operated by the Ohio Department of Natural Resources. (Courtesy of LESI.)

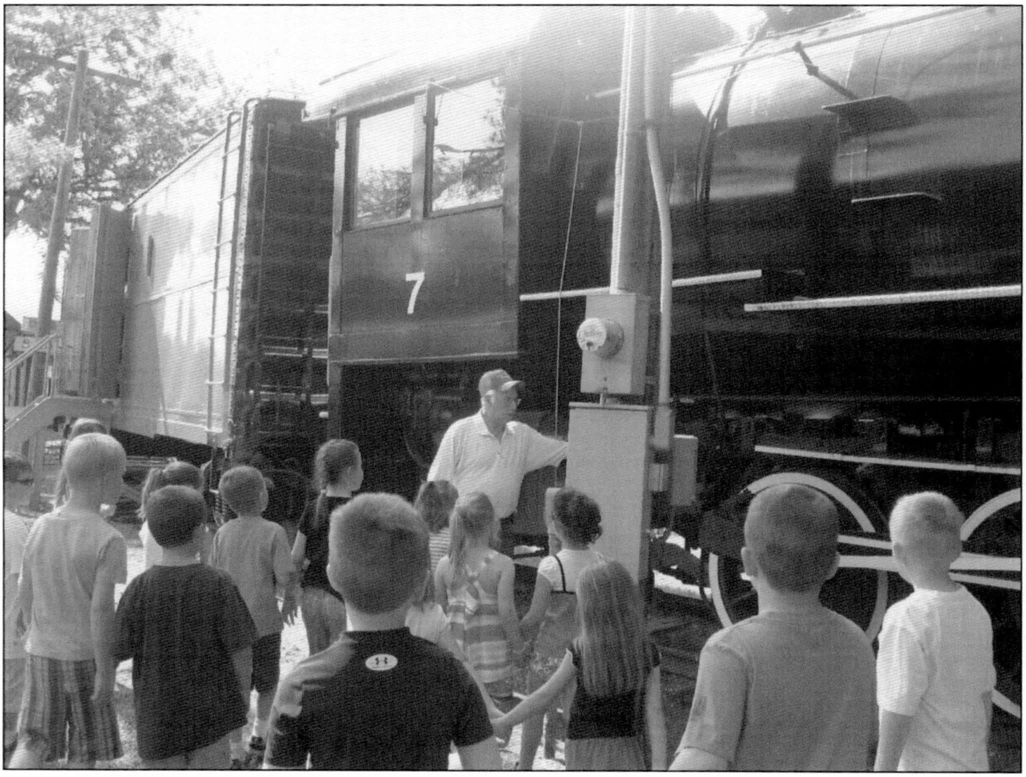

In 1976, the Mad River and Nickel Plate Railroad Society established a museum in Bellevue to give children of all ages the opportunity to climb aboard old trains. The first Ohio railroad was dedicated in Sandusky in 1835, and Bellevue has become a major railroad center. The museum is located on the property where railroad magnate Henry Flagler once lived. An early partner of John D. Rockefeller, Flagler went on to build the railroad that connected Key West to Miami. (Courtesy of Mad River and NKS.)

Named for Lyme Township, Ohio, Historic Lyme Village was started by area residents who sought to preserve the Historic Lyme Church. Over the years, the village, located at Ohio 4 and Ohio 113, has grown into a site of restored 19th-century buildings. Since 1976, the village, seen here around 1980, has hosted special events, building tours, and crafts days. (Courtesy of Historic Lyme Village.)

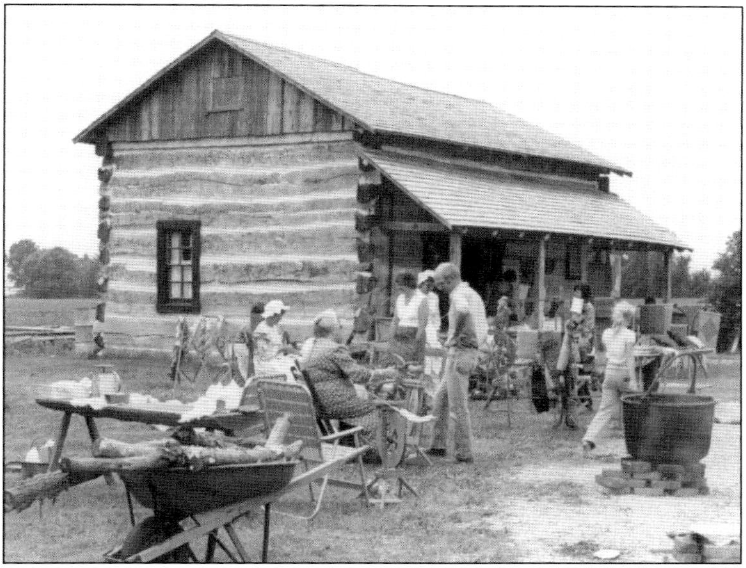

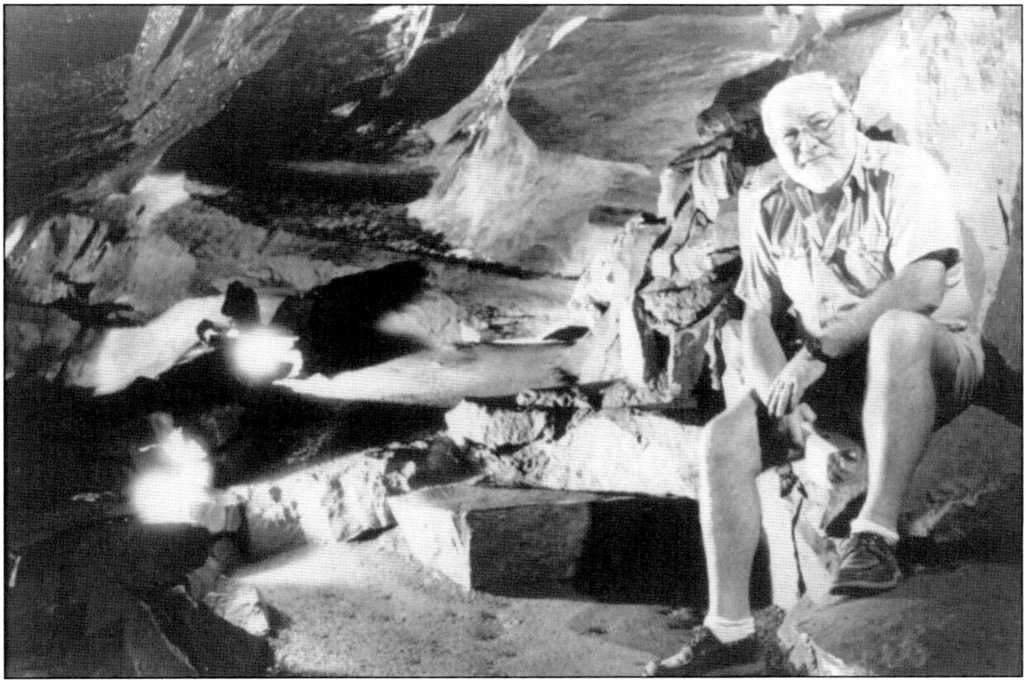

Since 1933, Seneca Caverns has drawn tourists to the underground mystery of a cave discovered in 1872. The cave includes a series of passageways and rooms, plus an underground river that feeds into the Castalia Blue Hole and, ultimately, into Sandusky Bay and Lake Erie. In this 1976 photograph, second-generation owner, the late Richard "Dick" Bell, poses deep inside the cavern. Bell was a founder of the Ohio Travel Council and a leader in regional, state, and national travel and tourism associations for many years. (Courtesy of Bell family archives.)

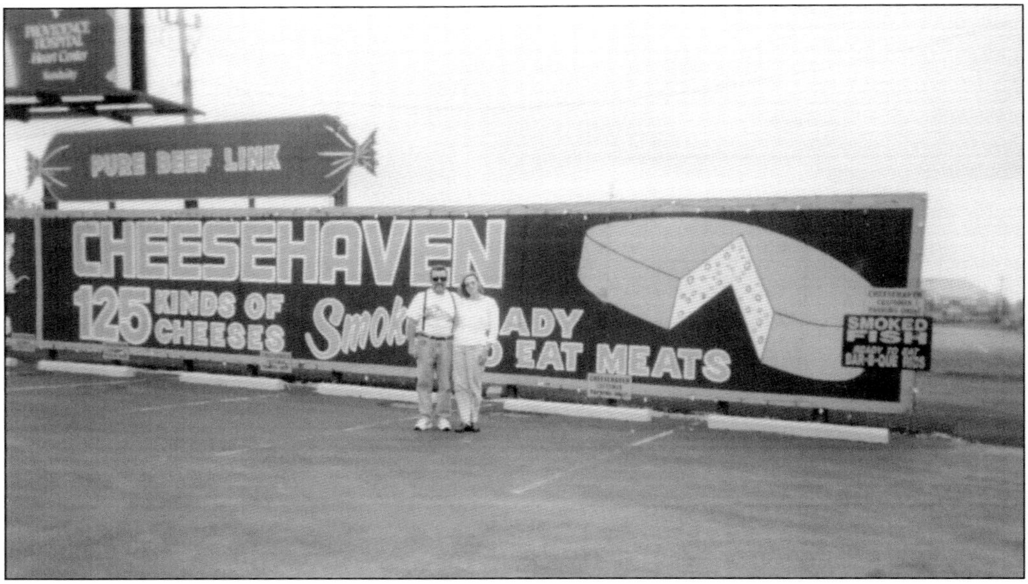

Richard Brassel started a family produce stand in 1949, which he expanded into a storefront to sell Swiss cheese. The family-owned business expanded over the years, and Cheesehaven became one of the most popular tourist businesses in the shores and islands region. The Cheesehaven lineup includes a world-class selection of cheeses in addition to island wines, candy, and smoked meat and fish. Pennsylvania tourists Ben and Missy Martilotta stand in front of the Cheesehaven sign around 1986. Their annual visit to the area always includes a stop at Cheesehaven. (Courtesy of the Martilotta family.)

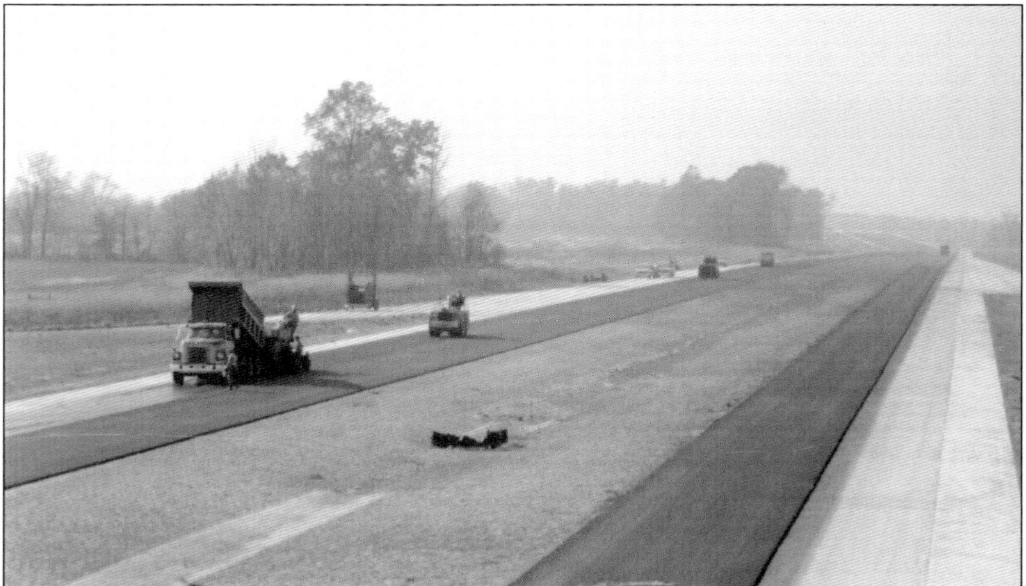

Local tourism and business interests won a hard-fought victory in 1986 with the opening of the 10-mile Ohio Route 2 Bypass around Huron. This addition to the area's highway infrastructure made access from Cleveland and other eastern markets much easier. Area visitors now had a four-lane road from Cleveland to Port Clinton, including the beautiful Thomas A. Edison Memorial Bridge over Sandusky Bay, which had opened in 1965. (Courtesy of Sheila Ehrhardt.)

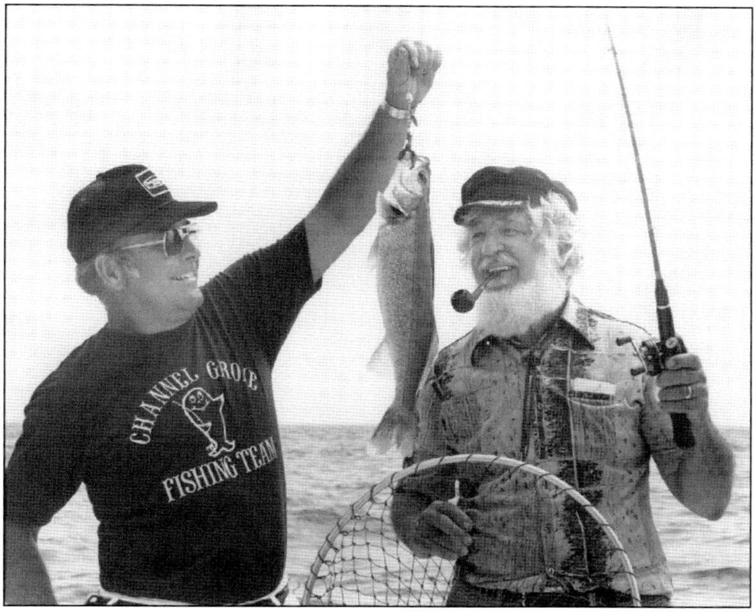

Channel Grove Marina owner Bob Brown (left) poses with fisherman Walt Siefert, showing off his first catch of the day around 1980. Sportfishing boomed in the late 1970s, as a cleaner lake allowed the walleye population to expand. The Ohio walleye harvest jumped from 112,000 in 1976 to more than five million by the mid-1980s. (Courtesy of the Brown family.)

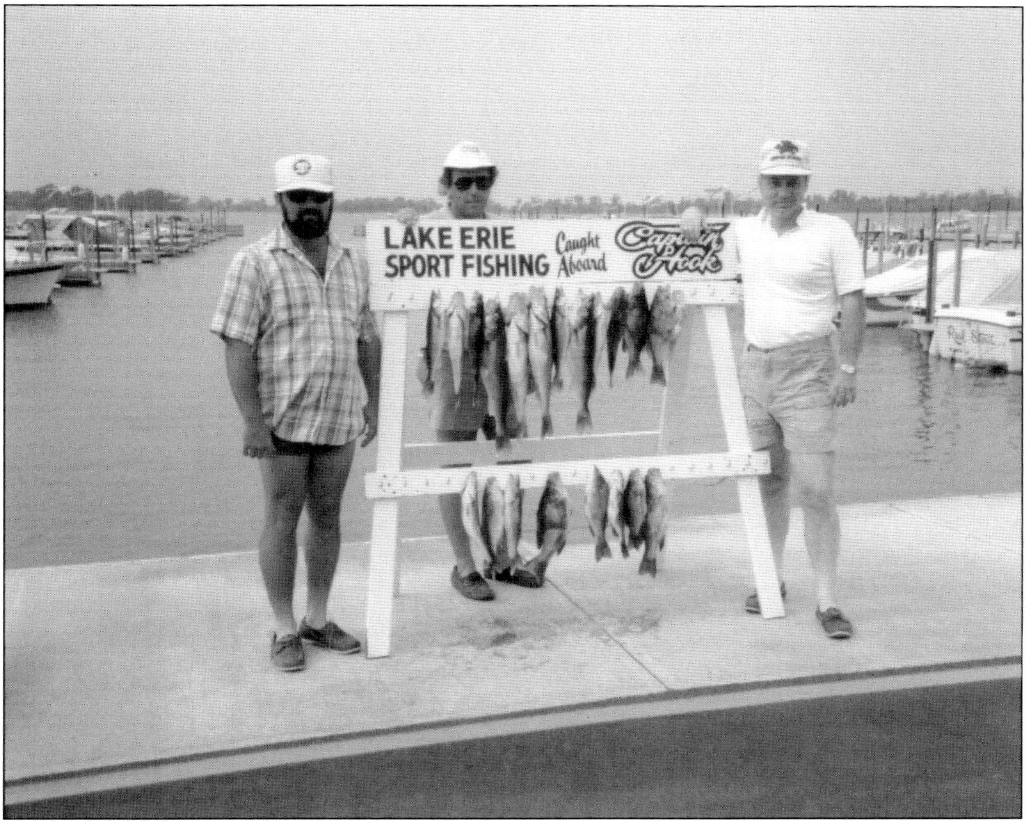

Fisherman Ben Martilotta (left), charter boat captain Pete Scheid (center), and fisherman John Hildebrandt pose for the camera in 1986 with their day's catch of walleye and smallmouth bass. A number of walleye fishing tournaments started in the area as the fishing improved. (Courtesy of the Martilotta family.)

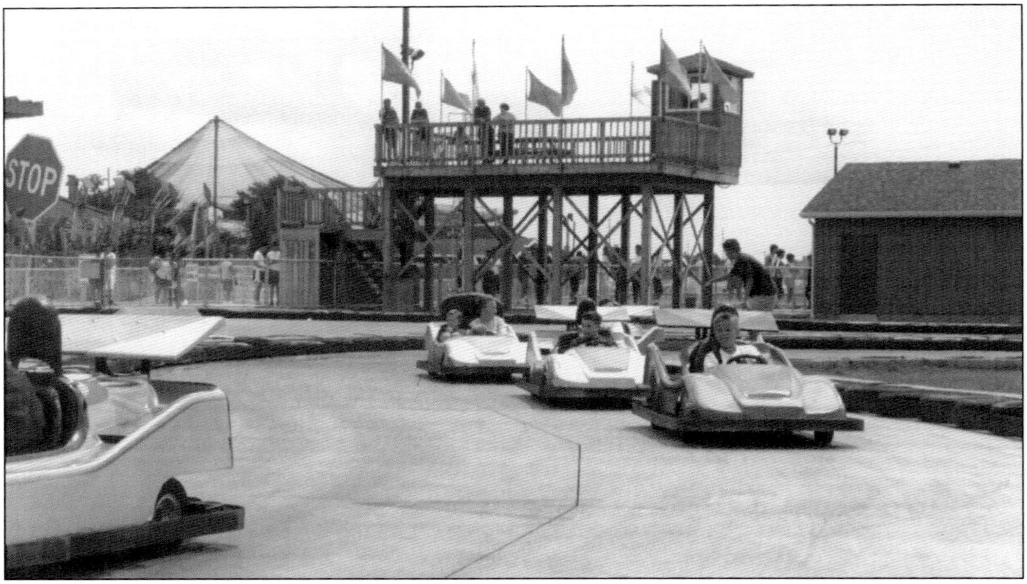

One of the first independent tourist attractions to locate on US 250, the main artery leading into Sandusky and Cedar Point, was Sports City, a family entertainment center with go-karts (seen here around 1990), arcade games, themed miniature golf, batting cages, and a gift shop. The attraction operated from 1989 until 2003, when it was removed for a new retail center. (Courtesy of Kula Hoty Lynch.)

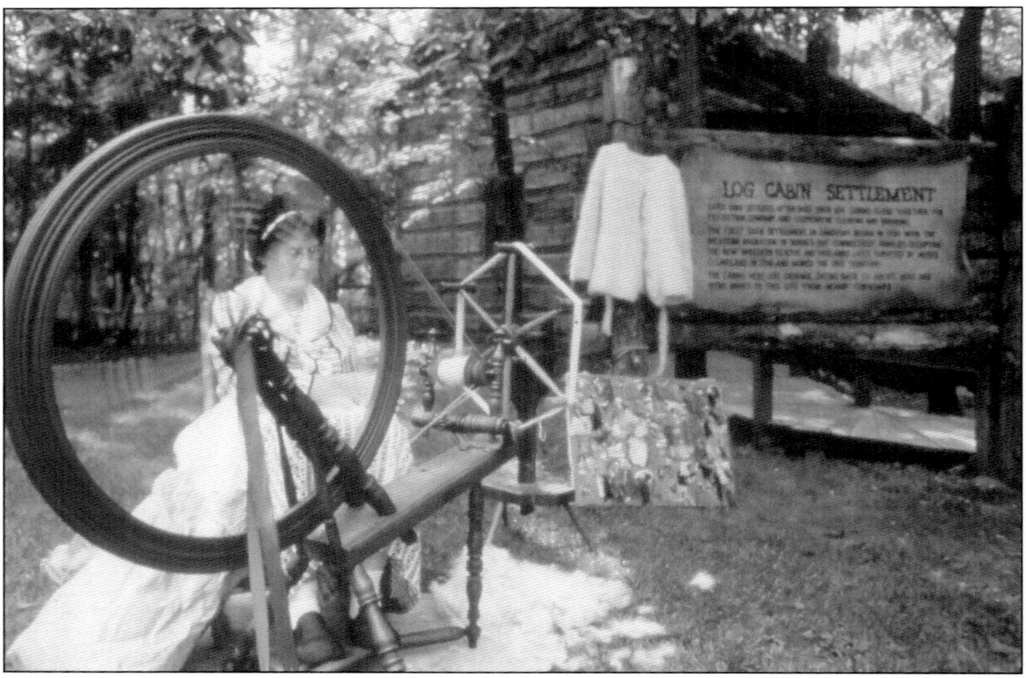

In 1971, the year following its centennial, Cedar Point introduced a new themed area, Frontier Trail. Craftspeople worked in authentic cabins and barns, introducing park guests to the crafts of yesteryear. Seen here about 1974, Granny Weatherall, a West Virginia craftsperson, spun dog hair into yarn for beautiful sweaters and coats, telling guests, "You're not spinning yarn if you're not spinning yarns." (Courtesy of CPA.)

The Corkscrew, which opened in 1976, was Cedar Point's first inverted roller coaster. It sent riders upside down three times, including a first-ever 360-degree vertical loop. Park guests were able to walk directly under the steel coaster, which spanned the midway. The popularity of the Corkscrew sparked an attendance boost for the park, which for the first time drew more than three million patrons. (Courtesy of CPA.)

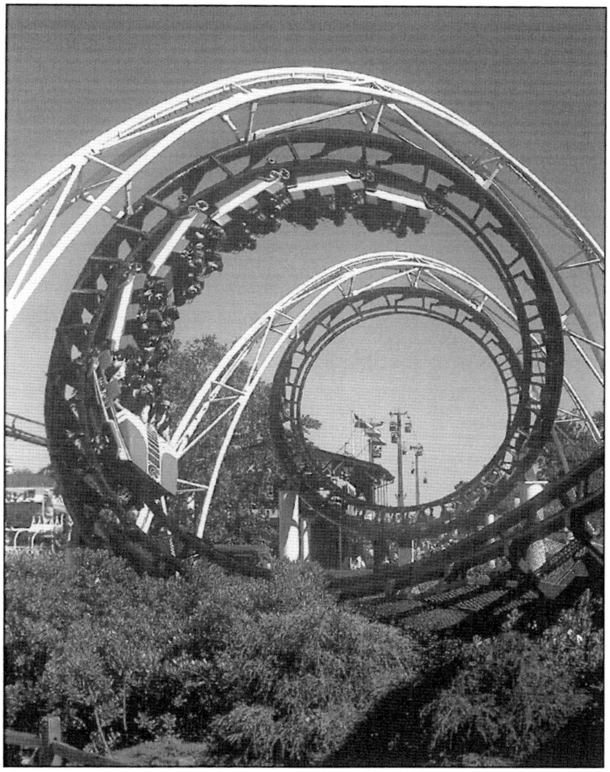

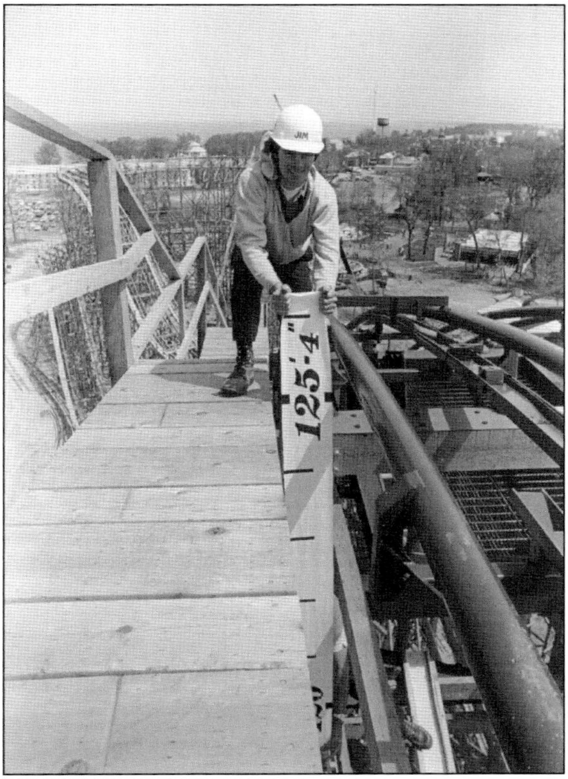

In 1978, Cedar Point opened Gemini, a twin racing coaster. In this Cedar Point publicity photograph, project manager Jim Colvin attests to the fact that the first hill of the Gemini tops 125 feet, making it the highest roller coaster in the world (it was also the steepest and the fastest). Gemini, even more popular than the Corkscrew, drove a new park attendance record in 1978. (Courtesy of CPA.)

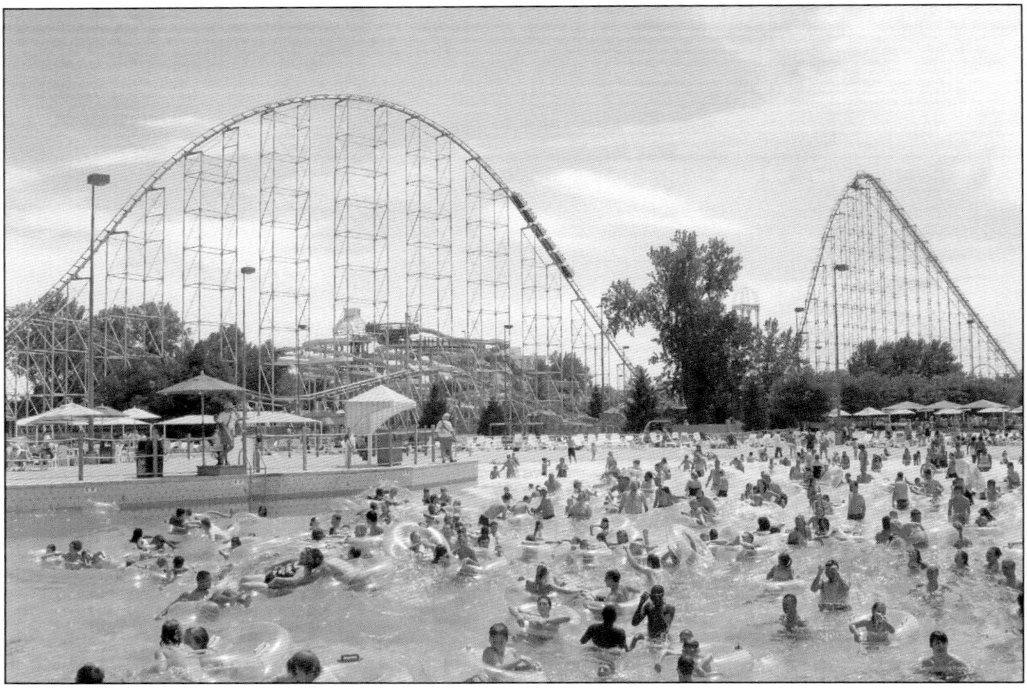

Cedar Point opened the area's first water park, Soak City, in 1988. The park has been expanded several times over the years. It is located next door to the amusement park, within sight of towering roller coasters, including the Magnum, which runs through the middle of the water park. (Courtesy of CPA.)

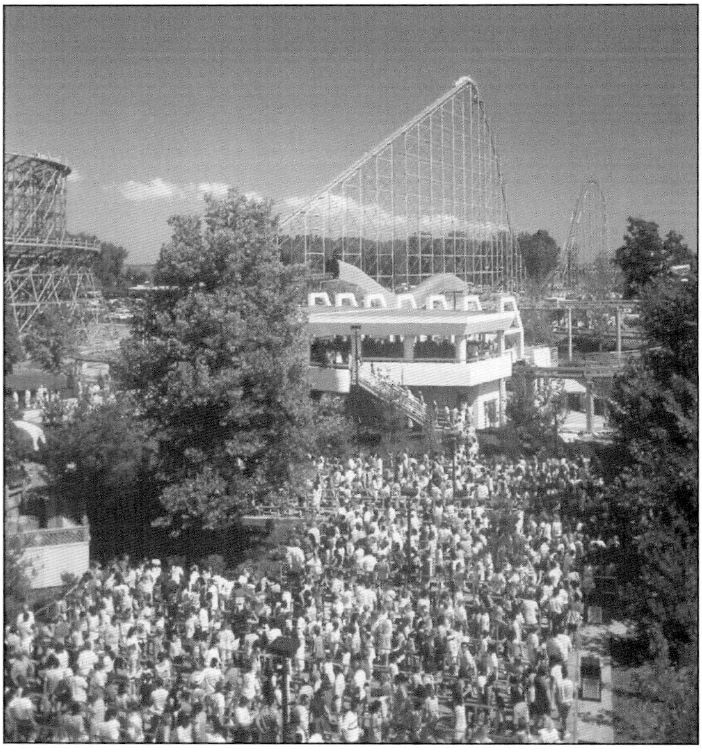

The Magnum XL-200, which opened in 1989, helped set a new park attendance record and sparked the start of the "coaster wars" within the amusement park industry. Magnum was the highest, steepest, and fastest coaster in the world when it opened; it was also the first coaster to break the 200-foot barrier, at 205 feet. It defined a new class of ride, called a "hypercoaster." From the top of the lift, riders are treated to a short, but incredible, view of Lake Erie, including Kelleys Island and the Marblehead Peninsula. On a clear day, riders can also see Middle Island and Pelee, both in Canada. (Courtesy of CPA.)

Seven

Roller Coasters, Water Parks, and the Jet Express 1991–2015

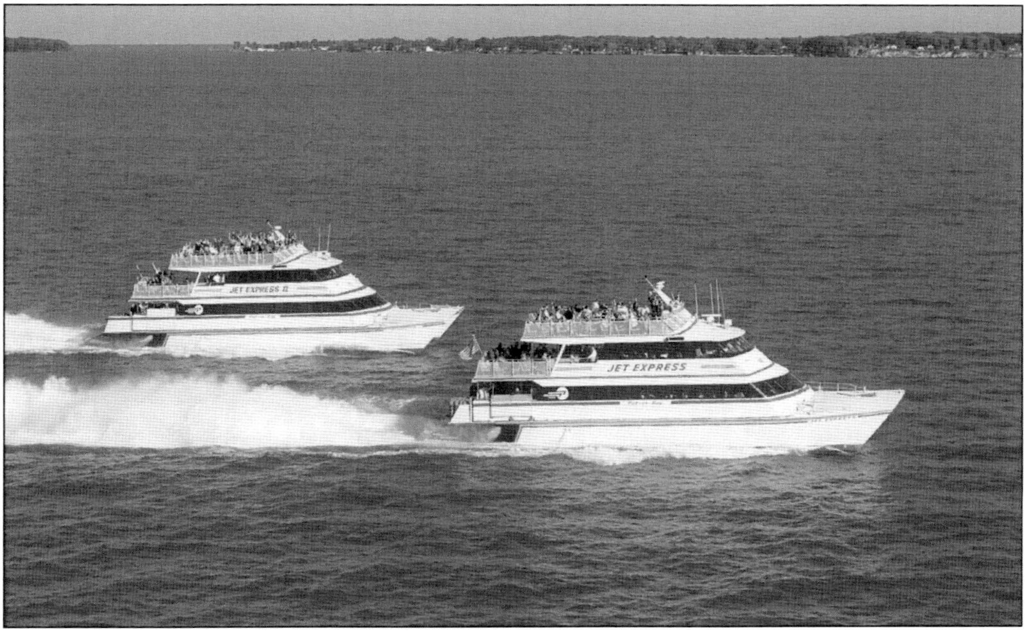

In 1989, the Put-in-Bay Boat Line Company abandoned its traditional freight and passenger ferry service from Port Clinton to South Bass Island, instead launching high-speed passenger service via a catamaran propelled by water jets capable of speeds up to 40 miles per hour. Travel time dropped from 90 minutes to 20 minutes. Businesses thrived as island visitation increased throughout the 1990s. A second Jet Express was added in 1992, and two more were added a few years later. The Jet Express, like the Great Lakes steamers of the previous century, has helped create a second golden age of tourism for the Lake Erie islands. (Courtesy of LESI.)

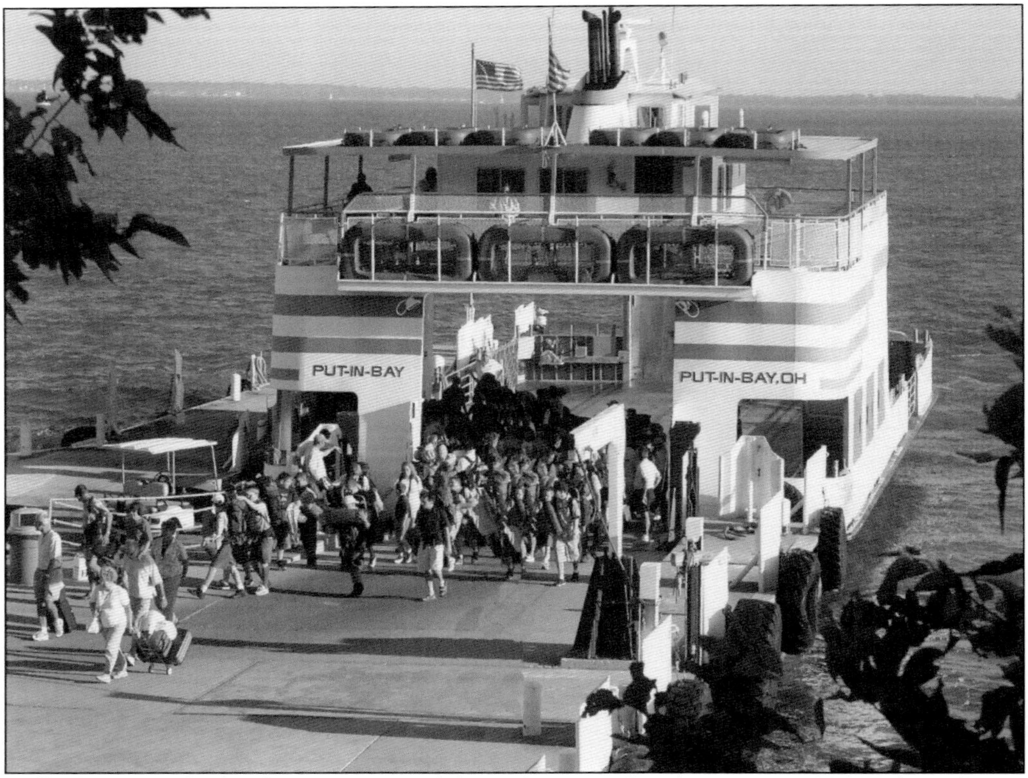

Miller Boat Line had its beginnings as a fishing charter and ice business in 1905. Today, it brings passengers, vehicles, and freight to South Bass and Middle Bass Islands from its base at the tip of the Catawba Peninsula on the mainland. Here, the ferry *Put-in-Bay* unloads vehicles and passengers at the Lime Kiln Dock at the southern tip of South Bass, just a three-mile run from Catawba. The Miller line, with its freight and vehicle capabilities, is the main transportation artery between the Bass islands and the mainland. (Courtesy of LESI.)

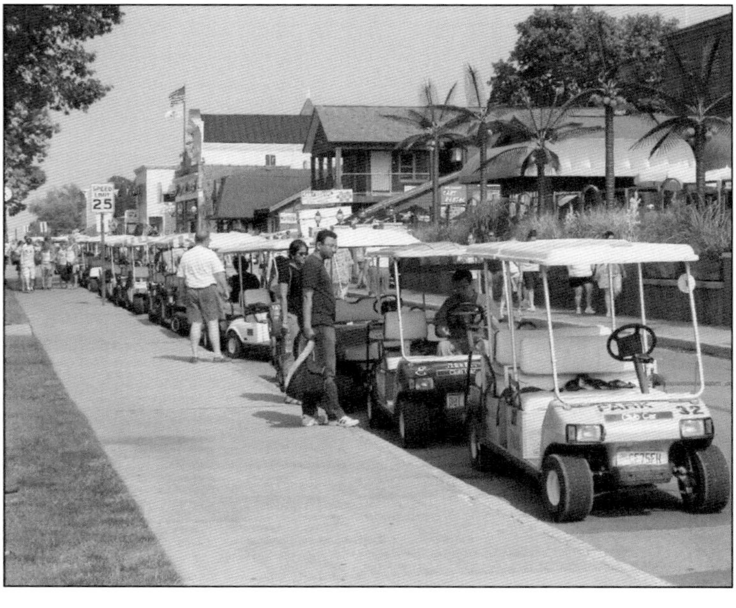

Golf carts have become one of the most popular modes of transportation on the Lake Erie islands. On most summer days, there are more golf carts on the islands than cars, as demonstrated here in downtown Put-in-Bay sometime between 2005 and 2010. While popular with tourists, and economically beneficial, some islanders lament the annual invasion of the golf carts. (Courtesy of LESI.)

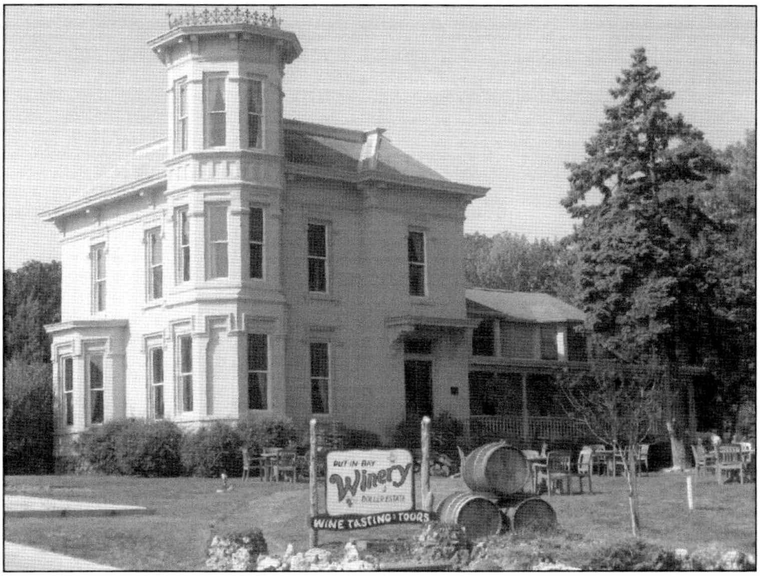

The historic Victorian and Italianate Doller Mansion was built by Valentine Doller between 1866 and 1885. A local entrepreneur, Doller also served as mayor of Put-in-Bay. The mansion was re-created as the Put-in-Bay Winery in 2009. (Courtesy of LESI.)

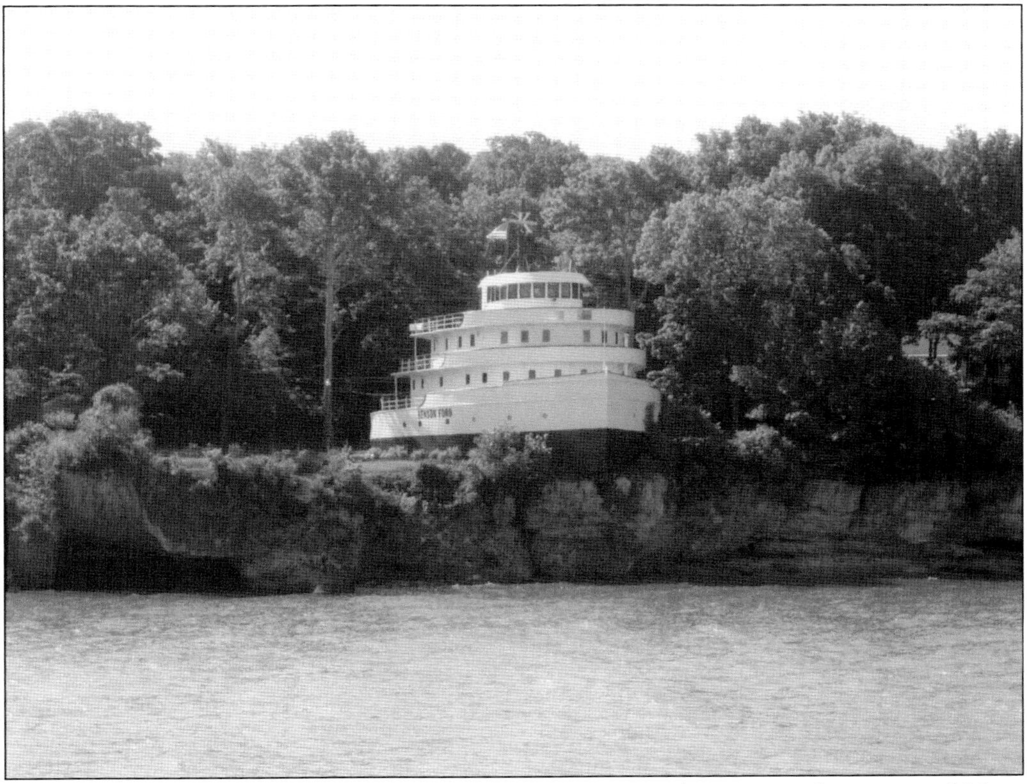

South Bass Island has long been a haven for summer retreats. This vacation home started life as the bow of the 612-foot-long lake freighter *Benson Ford*. Before the bow section was lifted into place on the west side of the island in 1986, the freighter hauled iron ore for the Ford Motor Company from Minnesota to Detroit for 50 years. Passengers on the Jet Express pass by the *Benson Ford* every day. (Courtesy of LESI.)

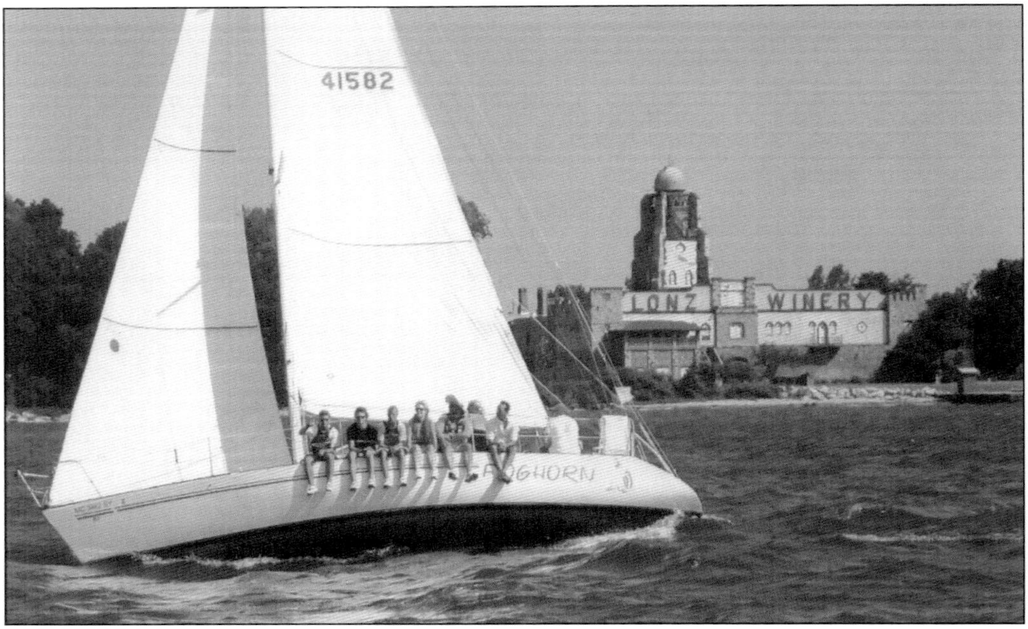

A sailboat passes the historic landmark of the Lonz Winery on the south shore of Middle Bass around 2010. Sailing continues to be a very popular recreational activity in the shores and islands area. The ILYA Regatta, held at Put-in-Bay, is still a major midsummer event, attracting boats and sailors from all over the Great Lakes. (Courtesy of LESI.)

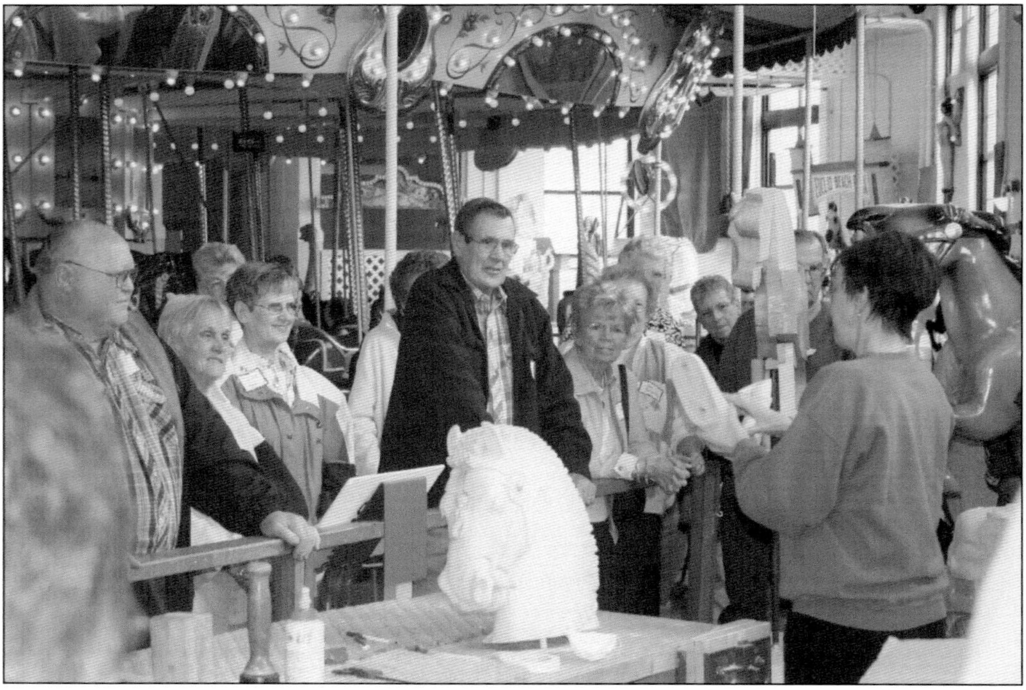

The Merry-Go-Round Museum in downtown Sandusky opened in 1990. Located in the former post office building in Sandusky, the museum is devoted to the art and history of carousels in America. The museum includes an operating carousel. Open year-round, it attracts visitors from across the United States. (Courtesy of LESI.)

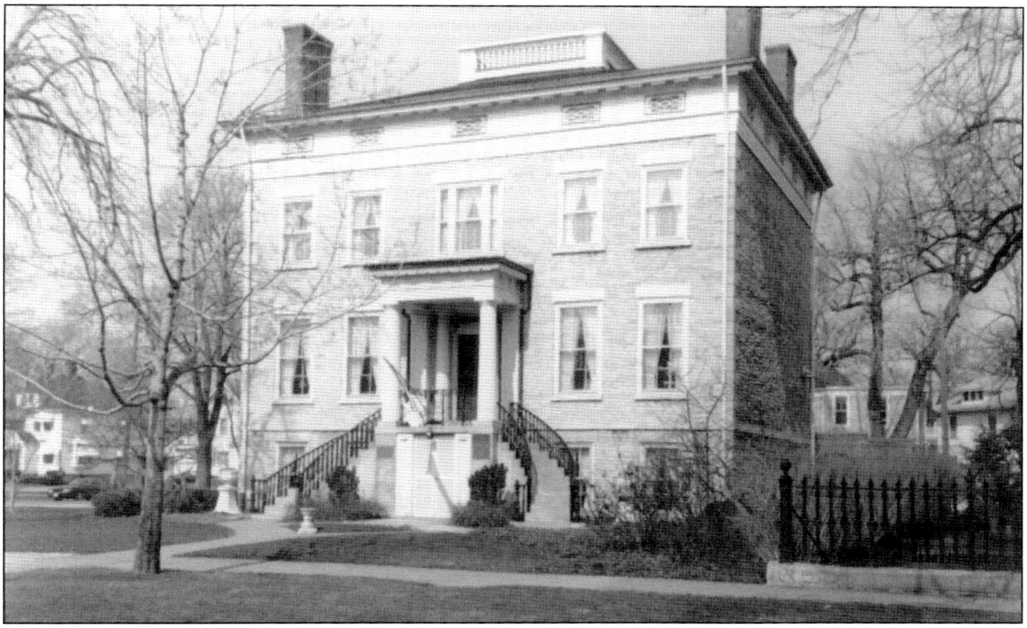

The Follett House Museum is located in a Greek Revival home built in 1837 for Oran Follett, a Sandusky businessman and a founder of the Republican Party. The museum features Native American artifacts and an extensive collection of objects related to the Confederate prison on nearby Johnson's Island. (Courtesy of SLARC.)

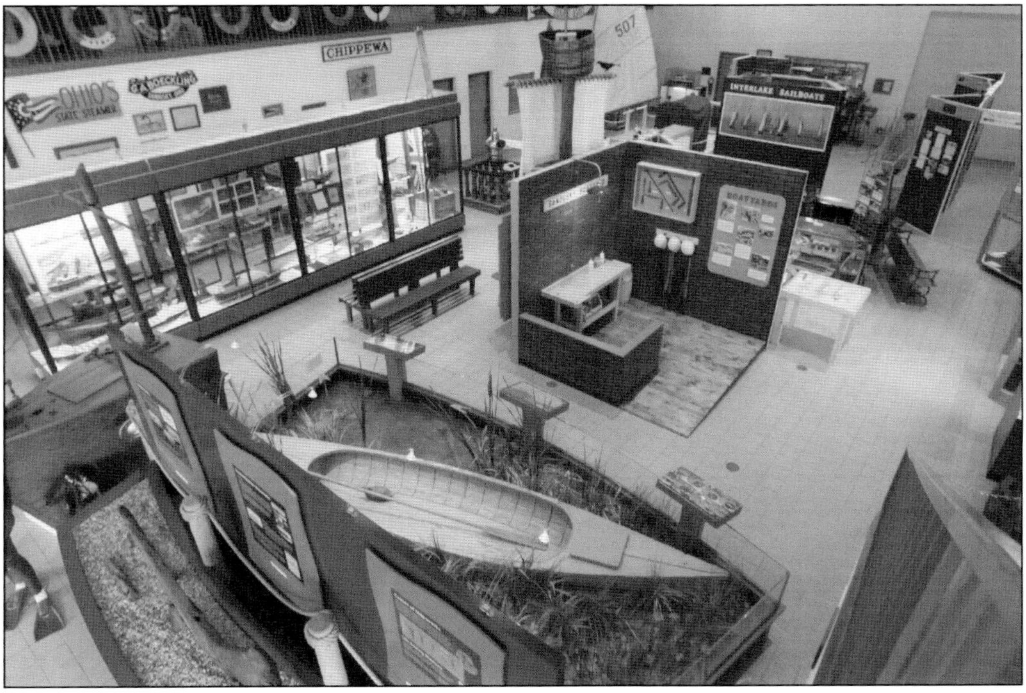

The Maritime Museum of Sandusky is dedicated to exploring the rich maritime history of the Sandusky area. As a major port on Lake Erie, Sandusky was a center for commercial fishing, coal shipping, and boat-building. The museum has also been involved in developing resources to educate visitors about Sandusky's significant role in the Underground Railroad. (Courtesy of LESI.)

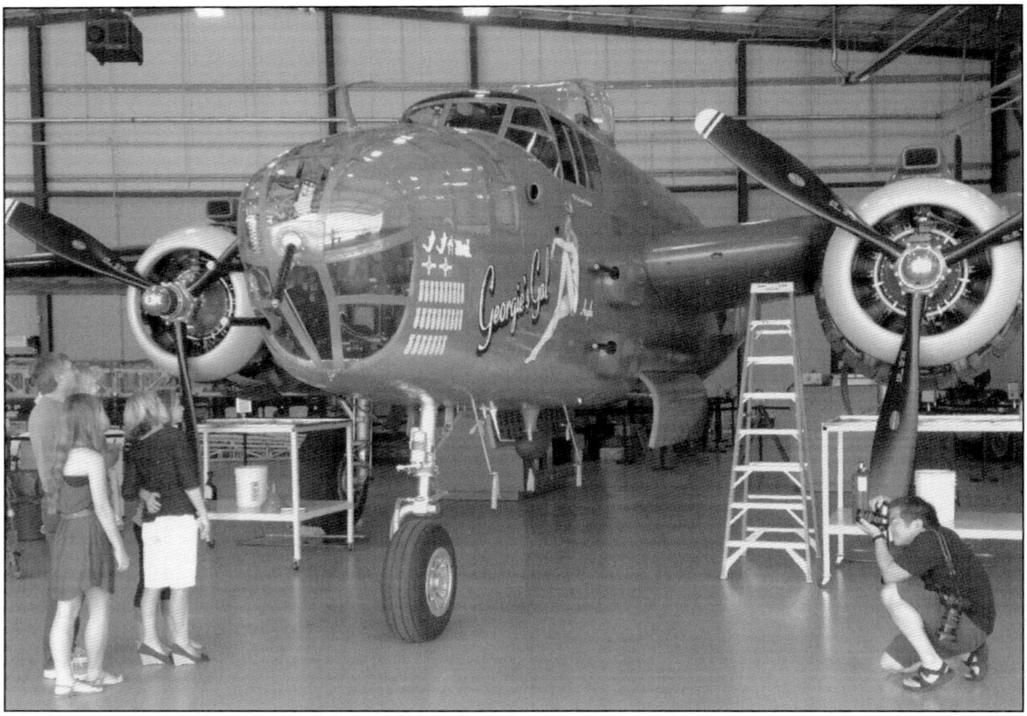

The newest museum to open its doors to area tourists is the Liberty Aviation Museum, adjacent to the Erie-Ottawa Regional Airport. These visitors pose in front of *Georgie's Girl,* a vintage World War II North American B-25 Mitchell Bomber. The museum, which opened in July 2012, also owns a flyable Ford Tri-Motor and is in the process of restoring a second plane. The museum also features a 1940s diner relocated from Jim Thorpe, Pennsylvania. (Courtesy of LESI.)

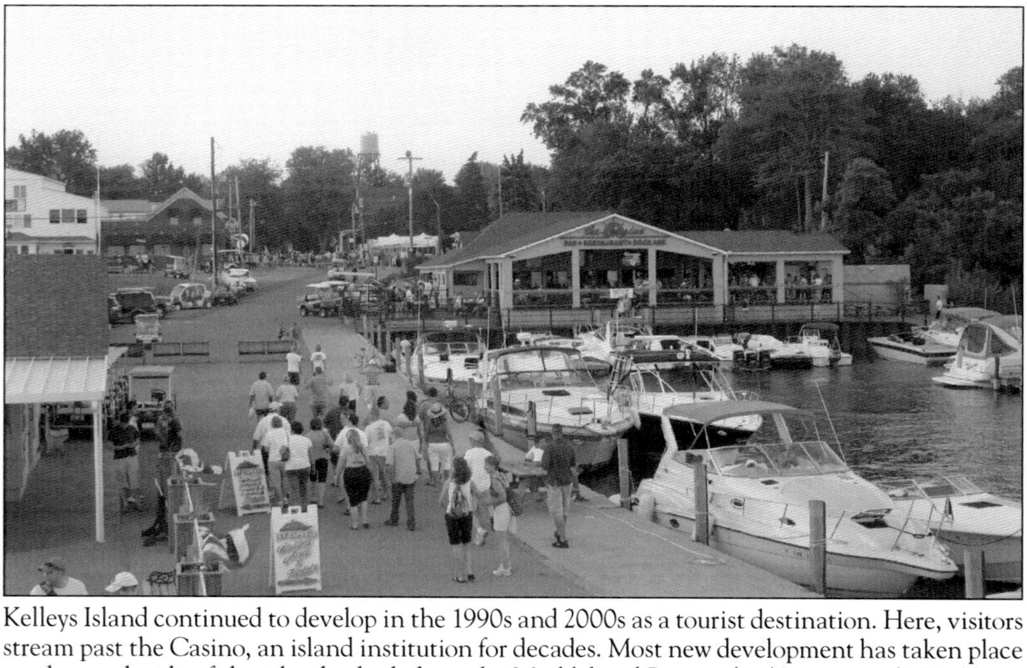

Kelleys Island continued to develop in the 1990s and 2000s as a tourist destination. Here, visitors stream past the Casino, an island institution for decades. Most new development has taken place on the south side of the island, which faces the Marblehead Peninsula. (Courtesy of LESI.)

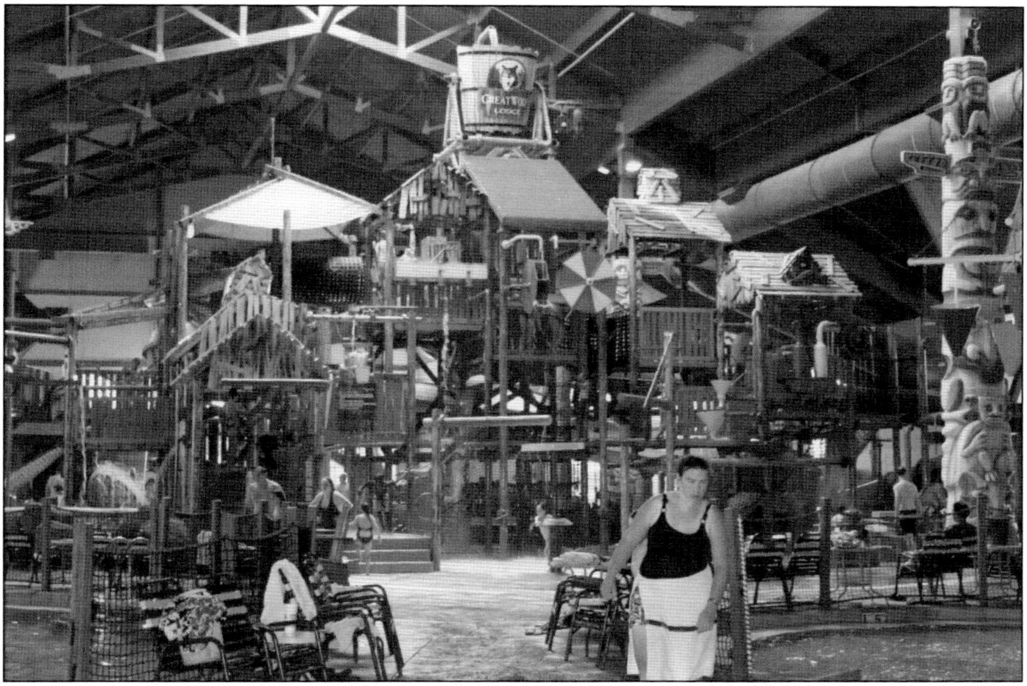

In 2001, a new era in tourism started in the region when Great Bear Lodge (soon to change its name to Great Wolf Lodge) opened on US 250 in Sandusky. Combining an indoor water park with a hotel, all in a North Woods theme, Great Wolf was an instant success and the prototype for a number of similar facilities. Open year-round, it brings young families—not just ice fishermen and ice boaters—to the area in the winter months. The concept of the indoor water park resort originated in the Wisconsin Dells. (Courtesy of LESI.)

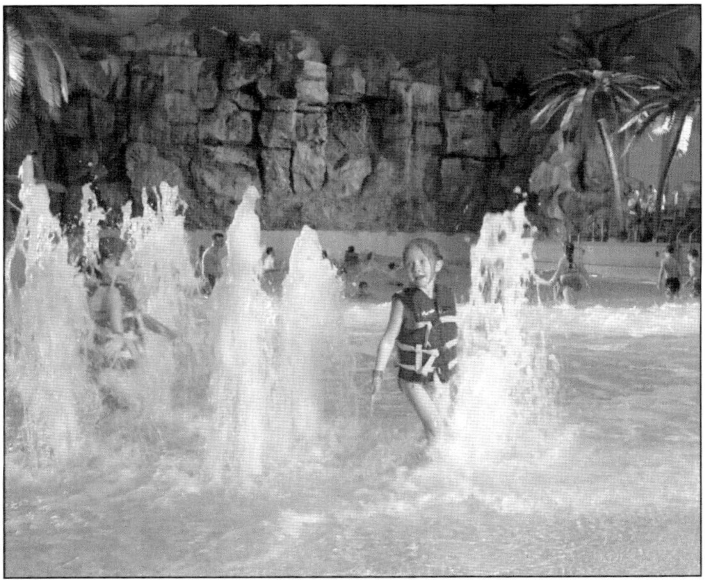

In 2004, Cedar Point opened Castaway Bay, a Caribbean-themed water park resort catering to younger families. Castaway Bay was a conversion of an existing Radisson hotel with the addition of a new water park. It was the first indoor water park in the region with an indoor wave pool. (Courtesy of CPA.)

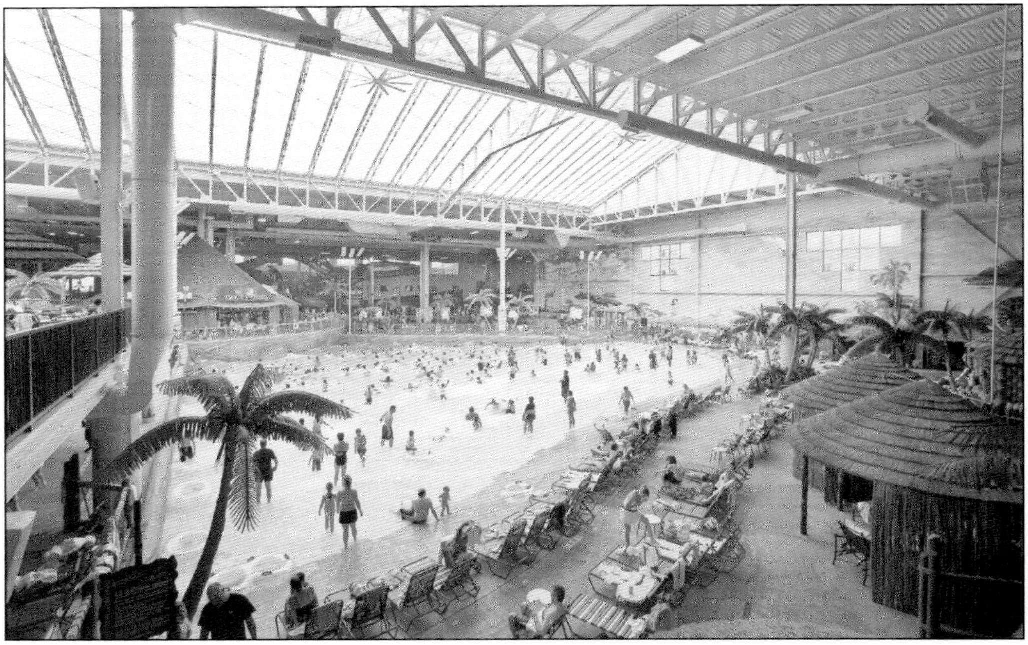

In 2005, Kalahari opened on US 250 just south of Sandusky. It is one of the largest indoor water park resorts in the world. Featuring an African theme, Kalahari has more than 800 hotel rooms and a water park with more than a dozen attractions covering 173,000 square feet. Kalahari also has outdoor water facilities and an animal park. In addition to the three major properties, several small water-park resorts opened in recent years, including Maui Sands and Rain. (Courtesy of Kalahari.)

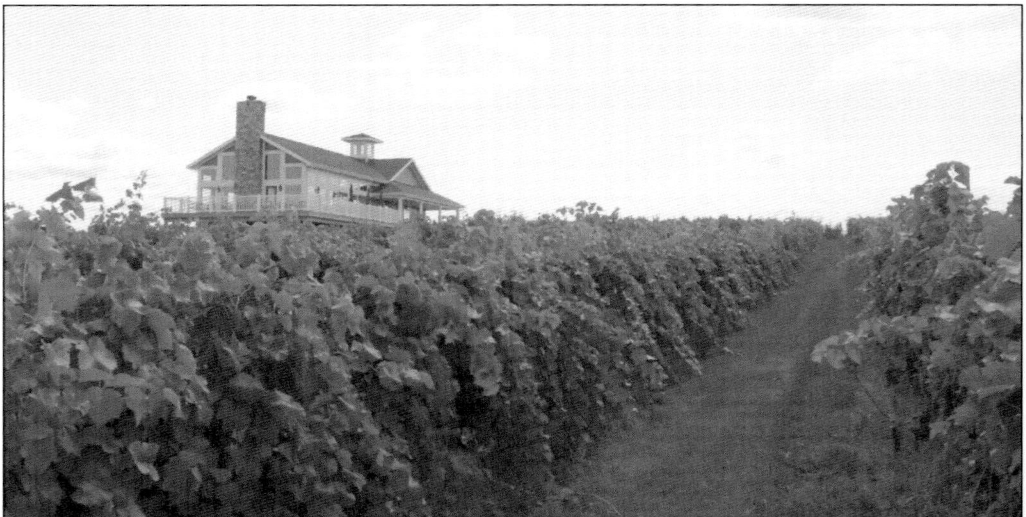

The number of small wineries in the region continued to grow in the 1980s, 1990s, and 2000s, reflecting both a state and national trend. In addition to the traditional sweeter wines made with Niagara and Delaware grapes, vintners began producing more varietal wines, like Cabernet and Chardonnay. The Quarry Hill Winery (pictured) opened in 2005. It is located near the highest point in Erie County and overlooks Edison Woods Metropark and farm fields. The blue line of Lake Erie can be seen a few miles to the north. (Courtesy of LESI.)

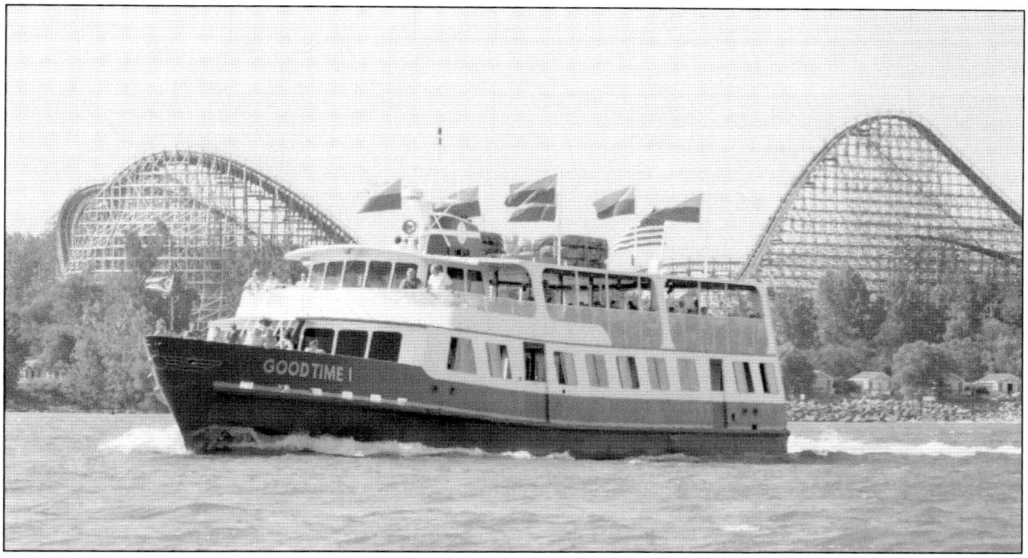

The *Goodtime I* has been offering Lake Erie cruises since the 1990s. The 117-foot-long, 315-passenger cruise boat docks in downtown Sandusky. The era of 300-foot-long lake steamers may have passed, but visitors still enjoy a good boat ride. In the early 1990s, Sandusky was also home to several other cruise boats—*City of Sandusky, Emerald Empress,* and the *Rocket*—but by 2000 the *Goodtime I* was the only cruise boat in port. In this photograph, Cedar Point's Mean Streak coaster and the cottages of Lighthouse Point form a striking background. (Courtesy of LESI.)

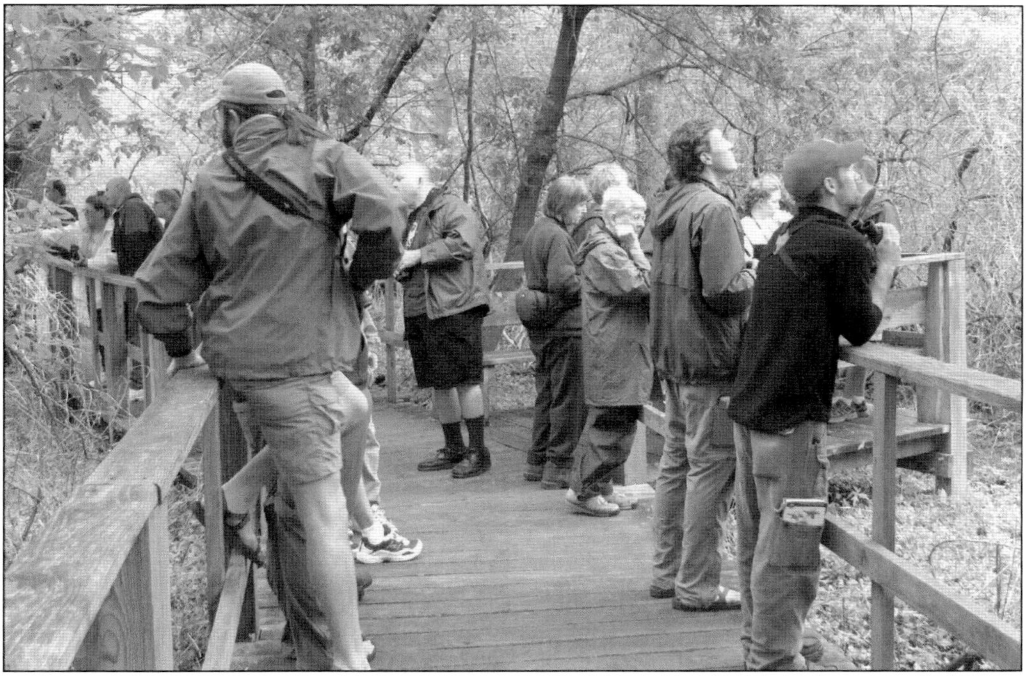

Birders flock to the Western Basin of Lake Erie, a major flyway for migrating birds, in the spring and fall. In the past decade, birding and other forms of ecotourism have grown significantly in the shores and islands region. A number of state parks and wildlife refuges along the southern shore of the lake provide prime spots for viewing migrating birds. The region also has a large bald eagle population. (Courtesy of LESI.)

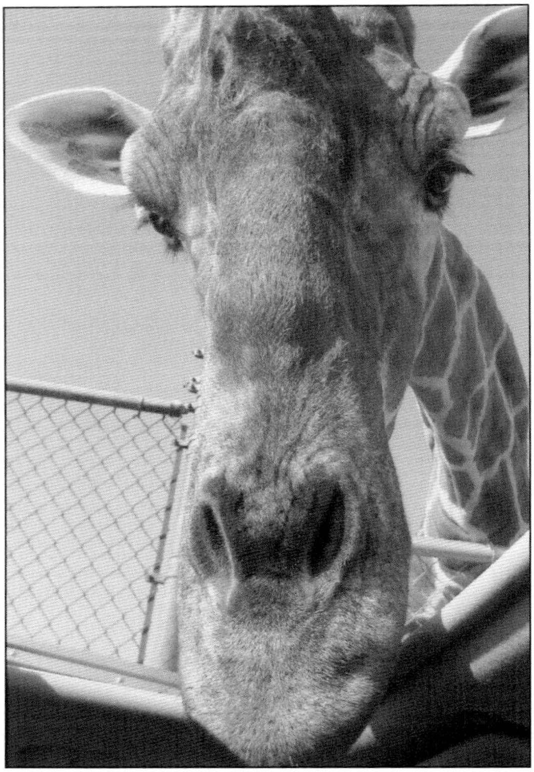

African Safari Wildlife Park on the Marblehead Peninsula has been a tourist attraction since 1969. Visitors have the opportunity to watch and feed animals from their cars as they follow a road through the 100-acre park. This giraffe is always hungry. The attraction also includes a small zoo. (Courtesy of the Boas family.)

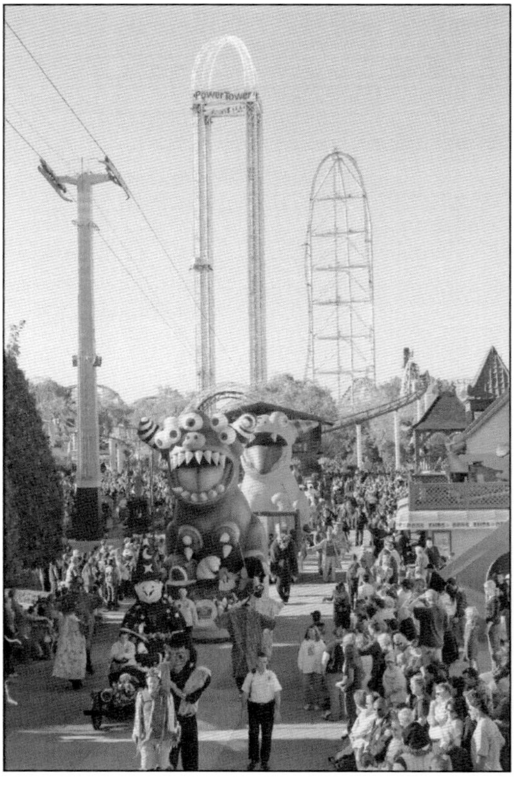

In 1997, Cedar Point introduced HalloWeekends, a Halloween-themed special event. HalloWeekends has been very successful for the park, extending the traditional park season by six weeks. One of the highlights of the event is the "Monster Midway Invasion Celebration," a family friendly Halloween parade down the main midway every Saturday and Sunday. At night, the action turns scary, with haunted houses and scare zones with roving monsters. (Courtesy of CPA.)

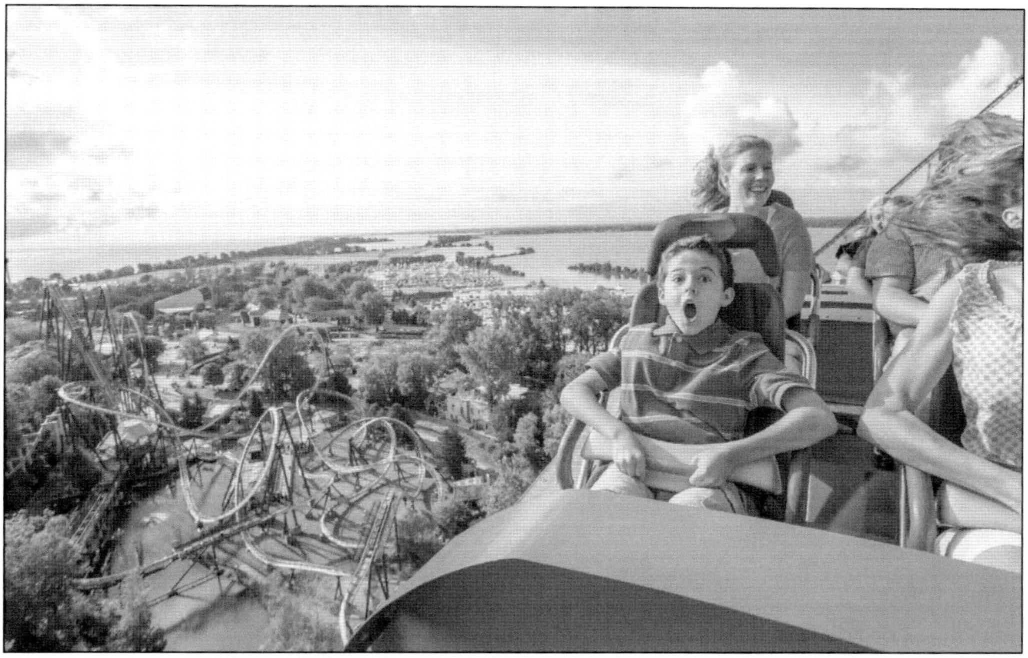

Cedar Point's growth in the past quarter century has largely been fueled by the introduction of world-class roller coasters. The coaster that defines the era is Millennium Force, which opened in 2000. It was the first coaster to break the 300-foot height level, but it is the total ride experience that makes the coaster so remarkable. The action is nonstop. Millennium has stood the test of time, as it has been consistently voted the best steel roller coaster in the world, including in 2014. This young rider has a white knuckle grip on the bar as the train crests the first hill. (Courtesy of CPA.)

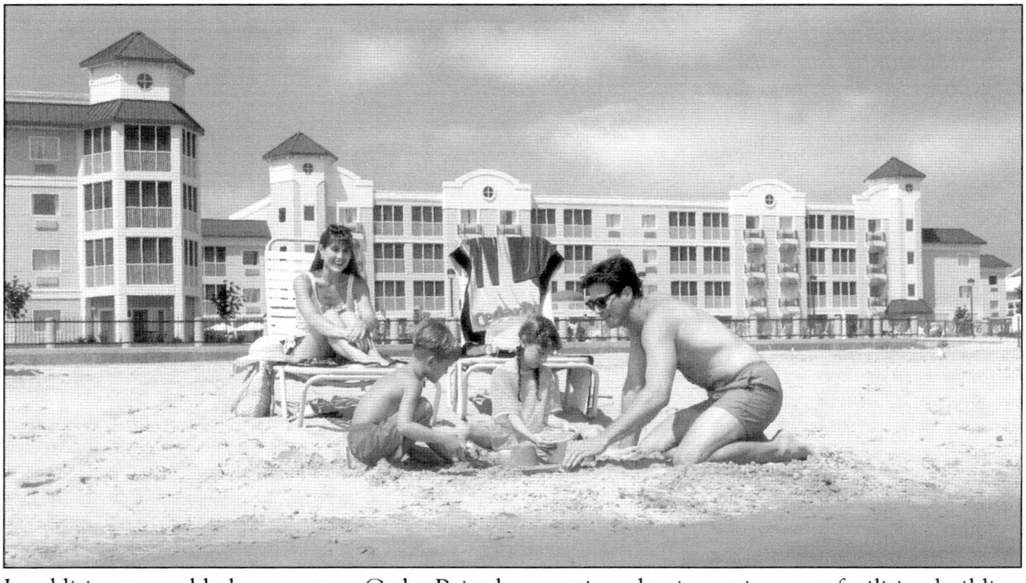

In addition to world-class coasters, Cedar Point has continued to invest in resort facilities, building several new hotels, including Sandcastle Suites (1990), Breakers East (1995), Breakers Tower (1999), Breakers Express (2000), and Lighthouse Point (2001). In 2015, Cedar Point is unveiling a complete refurbishment of the 1905 Hotel Breakers. This family enjoys Lake Erie and the sandy Cedar Point beach, much as families did 150 years before. (Courtesy CPA.)

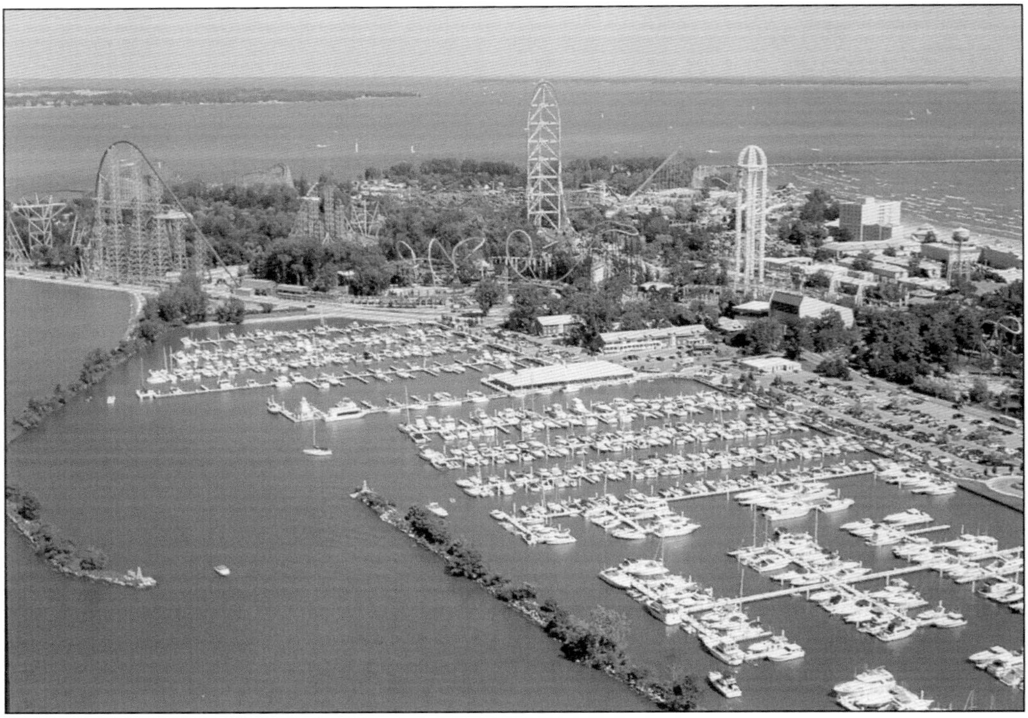

Cedar Point and Put-in-Bay continue to be the focal points for tourism in the Lake Erie shores and islands region, just as they were a century and more ago. In this aerial photograph of Cedar Point, taken around 2010, the park's relationship to Lake Erie is manifest. In the foreground is the marina, one of the largest on the Great Lakes. The beach is in front of the Breakers Tower. On the horizon are the Marblehead Peninsula, Kelleys Island, and the Bass islands. The roller coaster skyline, combined with the blue water of Lake Erie, compose an iconic view to three million park visitors every year. (Courtesy of CPA.)

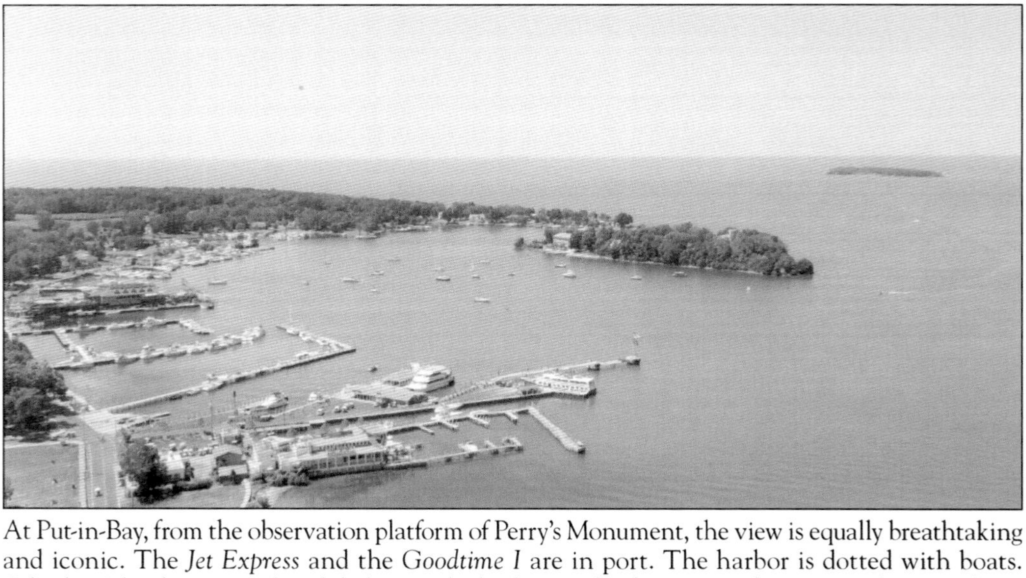

At Put-in-Bay, from the observation platform of Perry's Monument, the view is equally breathtaking and iconic. The *Jet Express* and the *Goodtime I* are in port. The harbor is dotted with boats. Gibraltar Island is a green hand sheltering the harbor. In the distance in this c. 2012 photograph are Rattlesnake Island and the blue horizon of Lake Erie. (Courtesy of the Hildebrandt collection.)

BIBLIOGRAPHY

Ayers, R. Wayne. *Ohio's Lake Erie Vacationland in Vintage Postcards.* Charleston, SC: Arcadia Publishing, 2000.

Davidson, Ron. *Sandusky, Ohio.* Charleston, SC: Arcadia Publishing, 2002.

Dodge, Robert J. *Isolated Splendor, Put-In-Bay and South Bass Island.* Hicksville, NY: Exposition Press, 1975.

Francis, David W. "Cedar Point and the Characteristics of the American Summer Resorts During the Gilded Age." *Hayes Historical Journal* (Winter 1988): 28–37.

———. "Steamship Service to Cedar Point, 1870–1852." *Inland Seas* (Summer 1977): 106–122.

———, and Diane DeMali. *Cedar Point.* Charleston, SC: Arcadia Publishing, 2004.

———, and Diane DeMali. *Cedar Point, The Queen of American Watering Places.* Sandusky, OH: Cedar Fair Entertainment Company, 2004.

——— and Diane DeMali. *Ohio's Amusement Parks in Vintage Postcards.* Charleston, SC: Arcadia Publishing, 2002.

Frohman, Charles E. *Cedar Point Yesterdays.* Columbus, OH: Ohio Historical Society, 1969.

———. *Put-In-Bay.* Columbus, OH: Ohio Historical Society, 1971.

Hildebrandt, Hugo John. "Cedar Point: A Park in Progress." *Journal of Popular Culture* (Summer 1991): 87–107.

Legibel, Ted, and Richard Wright. *Island Heritage: A Guided Tour to Lake Erie's Bass Islands.* Columbus, OH: Ohio State University Press, 1987.

Martin, Jessie A. *The Beginnings and Tales of the Lake Erie Islands.* Jessie A. Martin, 1990.

Ohio Writers Program. *Lake Erie Vacationland in Ohio, Revisiting a 1941 Travel Guide to the Sandusky Bay Region.* Huron, OH: March Fourth Publishing Company, 1999.

Waffen, Chad. *Ohio's Lake Erie Islands.* Bay Village, OH: Westfalia Publishing Group, 2006.

Witten, Sally Sue. *Port Clinton, the Peninsula, and the Bass Islands.* Charleston, SC: Arcadia Publishing, 2000.

Discover Thousands of Local History Books Featuring Millions of Vintage Images

Arcadia Publishing, the leading local history publisher in the United States, is committed to making history accessible and meaningful through publishing books that celebrate and preserve the heritage of America's people and places.

Find more books like this at
www.arcadiapublishing.com

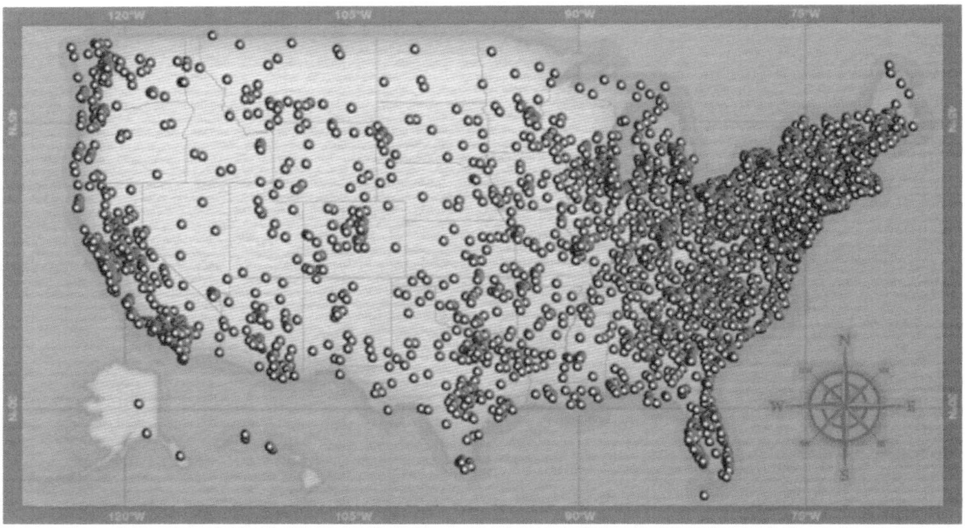

Search for your hometown history, your old stomping grounds, and even your favorite sports team.

Consistent with our mission to preserve history on a local level, this book was printed in South Carolina on American-made paper and manufactured entirely in the United States. Products carrying the accredited Forest Stewardship Council (FSC) label are printed on 100 percent FSC-certified paper.

MADE IN THE USA